Humanity and the Big Dipper

Humanity and the Big Dipper

*A History of Our Species
in Relation to Seven Stars*

Robert A. James

Figure 15 *reprinted by permission of Peanuts Worldwide LLC.*

Carlo Rovelli, The Order of Time *(2017) quoted by permission of Penguin Random House.*

ISBN: 978-1-7347846-0-2

Printed in the United States of America

Second printing

To the memory of my parents

CONTENTS

PROLOGUE

Ἄρκτόν θ', ἥν καὶ Ἄμαξαν ἐπίκλησιν καλέουσιν,
ἥ τ' αὐτοῦ στρέφεται καί τ' Ὠρίωνα δοκεύει,
οἴη δ' ἄμμορός ἐστι λοετρῶν Ὠκεανοῖο.

The Bear, which men also call the Wain,
Who circles forever in her place, shadowing Orion,
Alone not bathing in Ocean's stream.
  —HOMER, ILIAD

Help me Muse, I sing a song of patterns—the patterns our culture sees in a natural appearance, the stories our culture tells about those patterns, and the patterns and stories that other cultures see and tell. The song is both a celebration and a caution. A celebration, because we should treasure the striking similarities and differences across peoples. A caution, because much of what we have heard of the stories and patterns is simply not true.

The sky, like the Outer Space Treaty of 1967 calls the use of all of space, is "the province of all mankind." In settings remote in distance and time from urban lamps, and in places where and when people depended on the skies for navigation, agriculture and religion, the stars filled both our heavens and our imaginations. We moderns, by contrast and by and large, neither see all those points of light nor believe that our lives turn on a knowledge of them.

People in the Northern Hemisphere who know absolutely nothing else of stars—whether from astronomy, history or even astrology—still know one grouping above all others. The most prominent cultural relic of the Ice Age is the collection commonly referred to in the United States today as the Big Dipper. It is found in Stone Age carvings, cuneiform tablets, papyrus sheets, vellum pages, and legendary scrolls. It is usually the first natural pattern passed along or newly formed in one's life, dating to childhood immemorial.

This handful of lights is sufficiently bright, overhead, and set apart that it is often the only one recognizable from a metropolis. It is rarely seen by the uninitiated as part of any larger pattern. Unlike constellations with one or two bright points and a bunch of also-rans, it is composed of close peers. The Roman poet Manilius could thus uniquely hail the "seven equal stars."

The pattern instantly conveys to our eyes the unmistakable sense of sky. You know you are looking upwards when you see its image.

The Dutch artist Vincent Van Gogh set this spangled scene above the bank of the Rhône river *(see Figure 1)*. He wrote his brother Theo of "the starry sky painted at night. … On the aquamarine field of the sky the Great Bear is a sparkling green and pink, whose discreet paleness contrasts with the brutal gold of the [gaslight]. Two colourful figurines of lovers in the foreground." We feel the energy imparted by the stars to and through the couple on the promenade.

The pattern is close to the celestial pole around which stars are to us in apparent rotation throughout our nights and years. In northern skies the Big Dipper never sets beneath the local horizon, and thus is constantly visible— it is "circumpolar." As Homer says, for the Greeks it never bathes in Ocean's stream.

Figure 1. Vincent Van Gogh,
Starry Night Over the Rhône, 1888

Throughout many cultures this grouping within a constellation, or "asterism," is therefore a symbol of North itself. The Dipper and the north star form the entirety of the flag of the State of Alaska *(see Figure 2)*. A North Korean military unit uses the pattern not once but twice in its banner. It adorns the coat of arms of the royal Bernadotte house of Sweden.

Figure 2. Flag of the State of Alaska

Alternatively, the pattern is a symbol of constancy, as in the logo of the Iridium satellite telephone. The first space shuttle to launch after the 1986 explosion was mission STS-26 on *Discovery* in 1988, and the Big Dipper on the new mission's patch bore witness to NASA's commitment to keep flying in permanent memory of the loss of the seven *Challenger* astronauts.

The stories told about such a simple array of seven stars (plus a faint companion) form an important part of the lived experience of our species. But knowledge of our total treasure has been abdicated to the scholars.

The learning is concentrated in two sources. First, there are heavy works of astronomy gathering the stories about many asterisms and constellations—southern as well as northern, dim as well as bright, modern as well as ancient, obscure as well as famous. Second, there are heavy works of anthropology gathering the stories told by a single culture or close group of cultures about many kinds of natural occurrences and myths.

There are excellent books, but many are intended for fellow aficionados—scientists, cultural researchers, linguists, and other specialists. The literature often lies beyond the reach and interest of those who frankly know or care nothing about the stars—nothing, that is, except perhaps this pattern.

I had heard bits and pieces of stories spanning many cultures over the years, and repeated them—for example, to unsuspecting children and their parents on scouting campouts. To my charges, I spun these yarns:

- *The Big Dipper is the rear end of a larger constellation that has been known since prehistoric times as the Big Bear. If you look hard enough, you will see that bear, and her son, the Little Bear.*

- *The three stars at the end in a row look just like a bear's tail. Look closely at the middle one, and you will see a genuine double star. Using the two as a vision test, Native Americans first named them the Horse and the Rider.*

- *Since the Big Dipper is always in the northern sky and never sets or rises, it was especially valuable to everyone in olden times.*

- *Before the American Civil War, it was also called the Drinking Gourd. Countless souls escaping slavery would memorize the song "Follow the Drinking Gourd," its lyrics containing secret clues to the journey northward to freedom along the Underground Railroad.*

- *Look, this clear graphic I retrieved from the web* (see Figure 3) *shows exactly how the position at nightfall of the Big Dipper rotates with the seasons.*

Figure 3. The Internet Rotation, As Shown to Children

Years later, I realized I had not verified anything that I was saying to the kids. Sure enough, when I did the briefest of online research, I was embarrassed to find that *each of these tales is either flat-out wrong or seriously misleading.*

Red-faced, I set out quietly to correct my half-dozen mistakes, in an attempt at redemption. I found there is an immense amount of material behind the varying handfuls of anecdotes found in the reference works. One scientific discovery led me to a cultural revelation, then to a linguistic connection, then to a literary reference, and finally to this unexpected book.

This work is one layperson's attempt to rescue and retrieve from the vast academic literature just the leading stories that humans have told about this single pattern. It is a cameo of our collective imagination, as applied to these seven stars.

In my time and circles, we universally refer to the pattern as an implement—a household utensil, namely the Big Dipper. Any United States resident who calls it by a different name, or claims to see a larger "greater bear" object in the sky, or tells a story about four, seven or eight individuals, would be branded an outlier. In other times or circles, it is the Big Dipper that would be the curiosity. My goal is fully to open all our eyes that some cultures have made the same connection as we have, and that others have made completely different connections.

At the dawn of the West, Homer already knew that the pattern went by different names: "the Bear, which men also call the Wain." In book 18 of the *Iliad*, the starry scene of my epigraph is etched into the shield that the gods' blacksmith Hephaistos forges for the petulant warrior Achilles. In book 5 of the *Odyssey*, the same word-formula is used by the nymph Calypso to describe the constellations that Odysseus should keep on his left as he steers home to Ithaca, eastward from the island of Ogygia. (Alas, our hero experiences many twists and turns before he reunites with his wife and son.) Like Homer, we should know that one person's animal is another person's wagon.

This book could have been a bare itemization of the names that cultures have given to the pattern; indeed, that was my original intention, akin to a stamp collection. But I learned that animals, wagons, gourds, and dippers are just examples of only one type of pattern—a discrete object pictured from the apparent positions of the seven stars.

After sifting through the stories, I decided that it is not enough to set out the various object names, to recount simply that "tribe X sees a cat while tribe Y sees a bat"— though there is a large amount of *that* kind of recounting in what follows. By itself, such a tour would miss the larger gallery exhibiting the experience and creativity of our species. Moreover, the saga courses through the veins of titans of culture: Homer, Hesiod, Aristotle, Cicero, Vergil, Ovid, the Bible, Boethius, St. Jerome, Ælfred the Great, Geoffrey Chaucer, Dante Alighieri, Luis de Camões, Edmund Spenser, William Shakespeare, John Milton, Alexander Pope, Roger Williams, Cotton Mather, John Keats, Walter Scott, Nathaniel Hawthorne, Henry David Thoreau, Alfred, Lord Tennyson, Vincent Van Gogh, Federico García Lorca, and John Coltrane!

Peoples far removed from one another have seen a small number of patterns, from which distinct stories and types of stories have been imagined or set. All these patterns and the tales associated with them, not just those related historically or ethnically to our tribe and our time, accompany the stars themselves in what is in sum the common inheritance of all of humankind.

I am sure this volume contains far more information about the Big Dipper than you ever expected to digest. I will appreciate your kind indulgence, and by the last chapter, I hope the persistent reader will feel a little prouder to be a human being.

WHAT OUR SENSES SEE

First, let us face the facts that are presented to our naked eyes. No telescopes or binoculars are necessary or even allowed. On one side of the north celestial pole, the most visible stars look like *Figure 4,* the brighter ones being disproportionately larger in this diagram.

Nature does not connect these dots; we do. We need to start somewhere, so again like Homer, let us begin in the middle of things.

What Americans like me see most clearly in the middle of this picture of the northern sky is in fact a seven-star pattern, one that we call the Big Dipper. Extra credit goes to those who, from that Dipper, can already trace the routes to Polaris and Arcturus; more about them later.

Figure 4. Unlabeled Northern Stars

From here things get a bit obscure—that itself is evidence why the Big Dipper is singular nowadays. *Figure 5* connects the dots in the traditional Western way, with labels for the most significant asterisms and stars. Creative humans have indeed made out a fainter Little Dipper or Little Bear, with Polaris the end of its "handle" or "tail"; a kite shape now called Boötes the Herdsman or Wagoner, anchored by ruddy Arcturus,[1] and a backwards question mark forming the head of Leo the Lion (punctuated by bright Regulus).

---

[1] I have long been fascinated by the word Boötes. The accent mark on the ö is not an *umlaut* (the signal in German and other languages to change the sound of the vowel) but a *diaeresis* (the signal to pronounce each of two consecutive vowels independently); in the Greek, the two *o*'s are omicron (o) and omega (ω) respectively. Boötes, literally a cowherd, was also anciently named Arktophylax, the Bear-Watcher. A cowherd in French is *bouvier,* known on these shores as the maiden name of Jacqueline Kennedy Onassis.

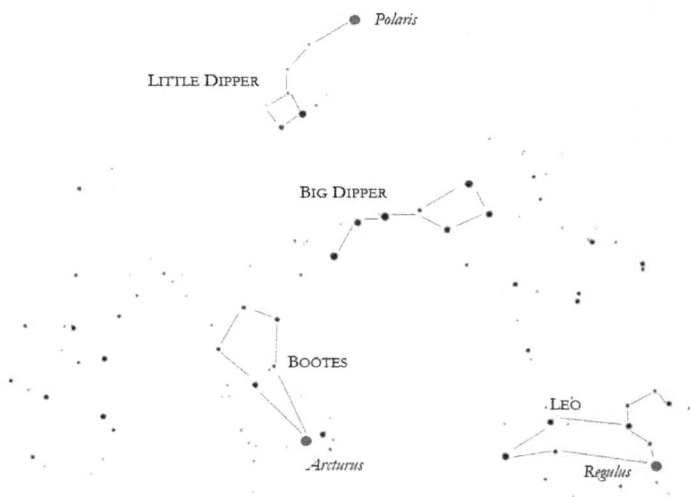

Figure 5. One Way of Labeling the Same Stars

The north celestial pole of the axis on which the Earth rotates presently lies in the direction of Polaris, which we moderns dub our "North Star." This star has not always enjoyed that distinction. The axis wobbles slightly or "precesses" over nearly 26,000 years, tracing a broad circle over our heads. The pole within recorded history has lain in the direction of other stars—one in Draco the Dragon, and elsewhere along the tail of the Little Bear. Many years ago, this part of the Little Bear was called the Dog's Tail, or Cynosura. Despite the upper tail's loss of navigational status, to this day a *cynosure* in English remains a center of attention or guiding principle.

Two thousand years from now, Polaris will have been supplanted as the best indicator by an ordinary star in the constellation Cepheus the King. But it is literally on top for our time, the only star that does not appear to us to move. Indeed, it will get closer and closer to true until the year 2095 CE. Let any pessimist, anyone who thinks everything is worse today than it has ever been, know that Polaris will be more and more reliably North throughout our lifetimes.

At first glance, Polaris itself is a mundane, unexceptional object. Astronomers would disagree, as I imagine they probably find every astral feature worthy of note. It is a double-star system, one of the first ever spotted. The larger of the two is itself a "Cepheid variable," a type of star with a consistent brightness cycle that the astronomer Henrietta Swan Leavitt showed us functions as a kind of yardstick for measuring the distance from us to its sisters in other galaxies.

Still, to laypersons, Polaris is nondescript, only about the forty-sixth most luminous point in the heavens. When it periodically dims, it even sinks to about number 50. Yet good old No. 46 is the most valuable star for navigation that we northerners have. It thus stands for the proposition that you need not be the best or the brightest if you can be in the right place at the right time. There is a lesson here for all of us.

Pity our fellow denizens down under, for whom the corresponding mark is unchaperoned by a South Star. Instead, they must extend imaginary lines from the Southern Cross constellation and from two other pointer stars, Alpha and Beta Centauri, to intersect somewhere in the distant blackness where the celestial pole must be. Northerners have a beacon; southerners get a high-school geometry problem.

Many other asterisms and constellations beyond this patch have been laid out by sailors and astronomers, astrologers and priests, poets and lovers. Besides Boötes and Leo, there are Corona the Crown; Orion the Hunter; Canis Major and Canis Minor the Greater and Lesser Dogs with their respective leading stars Sirius (literally the "scorcher") and Procyon (literally the "dog preceder"); and Taurus the Bull and its embedded clusters the Pleiades and the Hyades. Even for most northerners these are *not* circumpolar—they rise above and set below the horizon at specific times of the year. They have many stories of their own. Only some tales that connect them with the Dipper pattern will be told here.

Since the Big Dipper is so visible and spans such a great portion of the sky, it is a helpful reference to the rest of the northern heavens. First and as most generally known, the two end stars of the Dipper are "pointers" that show the way to Polaris (and thus to the Little Dipper). Proceed in a straight line about five times the apparent distance between the pointers and you are at true north, with the North Star patiently waiting to greet you. Second, the handle stars "arc to Arcturus," the bright red leader of Boötes (seen in *Figure 5*), and from there they "spike to Spica," the bright white champion of the constellation Virgo the Virgin.

Wait, wait, there is much more for those who care. The bottom of the Dipper "leaks to Leo" and its principal star Regulus (also seen in *Figure 5*), while the top "caps to Capella," brightest point in the constellation Auriga the Charioteer. Crossing the bowl diagonally leads to the leading light Castor in Gemini the Twins. Once you spot at least two of Virgo, Gemini, Leo and Taurus, constellations in the ancient Zodiac, you can trace the elliptic, the line of the plane near which the planets can be seen. And so on. Suffice it to say that the Dipper is handy.

But already we are headed dangerously in the direction of the astronomy books that describe *all* the stars and *all* the patterns and stories. Reader, I doubt you really care about Auriga the Charioteer, and I sense you may just now be nodding off. I intend henceforth to stick to the Dipper.

HUMANITY AND THE BIG DIPPER

WHAT IS REALLY THERE

Let us briefly turn to astronomical facts, mostly unknown to older cultures and indeed unknown to or forgotten by most of us today. The modern scientific convention is to divide the entire observed sky, not just the prominent asterisms, into 88 blocky regions named for their principal constellations. The Dipper pattern lies within Ursa Major, the Big Bear—or literally from the Latin, the "Greater She-Bear."

Attention must now be paid to an eighth star in the pattern—Alcor, sometimes called "the Rider." It lies adjacent to the middle star in the handle or tail—Mizar, sometimes called "the Horse." Alcor is formally known as *80 Ursae Majoris*—more or less the eightieth brightest star in the region.

Contrary to what I told the kids, Alcor and Mizar do *not* constitute a double-star system. For most of my life, they have been thought to be an optical double, about three light-years apart, with their shinings just happening to line up in our sight. In fact, each of Alcor and Mizar in turn is a true double star, and Mizar's double stars are in turn themselves both true double stars. But in 2009, researchers showed that collectively these six stars, which we see as two points, are in gravitational congress. So the duo are still not double stars, but collectively the sextuplet forms a connected system after all.

Alcor was perhaps the faintest of objects identified by the ancients. It holds outsized cultural importance for its magnitude. It pops into and out of the various stories, so I often refer to the Big Dipper as a 7+1 asterism. A dimmer star that sits next to a brighter one, amid other similarly bright lights, enriches possibilities for the stories that humans tell. Many authors can make use of something or someone that is diminutive among several larger peers, as we will see.

The modern conventional names of the stars come to us from medieval Arabian astronomy, enhanced from classical sources while European astronomy largely slept. The universe picture of the classical world was recapitulated in the *Mathematike Syntaxis* of Ptolemy of Alexandria, set down around 150 CE. It became known as Greek and Latin as "the great treatise," a brand name that was translated into Arabic as *Al-magisti*, and was preserved in the Arab world under its better-known name *Almagest*. Adorned with replacement Arabian star names, it was re-exported in that form westward in the Middle Ages.

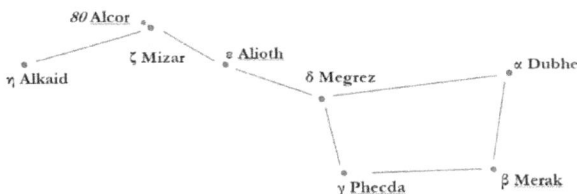

Figure 6. The Big Dipper with Arabian Star Names

The common translations shown in *Figure 6* and *Table 1* reveal a mixed brew among the story groups we will hear of below—parts of a bear, parts of a sheep, clothing of a mourner, and a mourner herself. The Bayer designation system usually ranks stars in order of apparent magnitude, but several exceptions were made. One was for this constellation, where Alpha (*a*) through Eta (*η*) Ursae Majoris are the Big Dipper lights in "pattern order," regardless of brightness. They are followed by the other points in Ursa Major (including Alcor) in rough order of brightness.

Table 1. The 7+1 Stars: Animal, Clothing, or Mourners?

Ursae Majoris	Arabian name (and common translation)	Apparent magnitude	Distance (light-years)
Alpha α	*Dubhe ("the bear")*	*1.8*	123
Beta β	*Merak ("the loin")*	2.4	80
Gamma γ	*Phecda or Phaed ("the thigh")*	2.4	83
Delta δ	*Megrez ("the base" or "the root" of the tail)*	3.3	81
Epsilon ε	*Alioth ("the tail" of the Eastern Sheep)*	1.8	81
Zeta ζ	*Mizar ("the girdle" or "the waistband")*	2.2	83
Eta η	*Alkaid ("the chief [of the mourning maidens]"), or Benetnasch ("the daughters of the funeral bier")*	1.9	101
80	*Alcor ("the faint one"), Al Suha ("the forgotten one"), Al Sadak ("the test"), or Al Saidak ("the true one")*	4.0	81

We immediately learn that a "test" of vision, being able to distinguish the "faint" or "forgotten" Alcor from Mizar, was prevalent in older cultures. The Horse and Rider names occur in German folklore, for example, and found their way into a star chart published in 1524. These terms therefore originated long before the attribution of the Horse and Rider names by Native Americans—who did not see modern horses until the sixteenth century CE. It is likely that they received the Horse and Rider labels from the same Europeans who brought the horses. Some observers have sniffed that it is not all that hard of a vision test, either—about a 20/20 task.

Here is some useful and genuine Arabian astronomical humor. If you have just done something really hard, while someone else has done something easy and wants equivalent credit, you can say "I show him al-Suha (Alcor)—and he shows me the moon." *(Rimshot.)*

The middle five (+1) of the seven (+1) Dipper stars are in a large physical star cluster, sometimes called the Ursa Major Moving Group (or, more prosaically, Collinder 285), such that they travel together. This cluster is distinct from the more distant end stars in our view, Dubhe out on the pointers end and Alkaid out on the handle end.

Over tens of thousands of years, the cluster will move one way, those end stars will move different ways, and the Dipper will flatten out and eventually flip. (In fact, the constellations have already moved some distance from how the astrologers have fixed them for horoscopes.) We humans live in contingent time, when these stars happen to present themselves where they were decades ago, radiating the light that has traveled so far before colliding with the backs of our eyes. Our ancestors, the story-makers and the story-tellers, did not know that.

WHAT OUR MINDS CREATE

Enough of star-gazing and astronomy. It is time to trace pictures and tell tales.

Are the patterns inherent in our brains? Some may believe that the connections we humans impose on nature emerge out of Jungian archetypes and our collective unconscious. Others may regard them as keys to a comprehensive assessment of one's psyche, akin to a Rorschach test. Still others may think they comprise a Gestalt summation of individual elements.

A second group of investigators has resisted these universal explanations, noting that ethnic lineages and historical migrations have produced similar stories only in some places and times, and not in others. Hence, specialists in anthropology, archaeology, linguistics, and other fields have written extensively on the meanings of the names attributed to these stars by historically, ethnically, or genetically related peoples. The relevant work is carried out in specialties within specialties, such as ethno-astronomy and archaeo-astronomy.

As a layperson, I necessarily take a different tack. A third and admittedly superficial approach, which I have adopted here, is to gather similar patterns as they have been seen by any of us, more or less oblivious both to the universalist impulses and to the social sciences. Humans have seen the asterism in several broadly discrete ways, which I have grouped into pattern categories *(see Table 2)*.

Table 2. The Pattern Categories of the Stories

Pattern Category One	*SCENE (Cosmic Hunt, Funeral Procession)*
Pattern Category Two	*ANIMAL (Bear or Elk, or part of an animal)*
Pattern Category Three	*IMPLEMENT (Wagon, Dipper, Plough, Drinking Gourd)*
Pattern Category Four	*ROLL-CALL (Seven Sages, Brothers, Sisters, Scoundrels)*
Pattern Category Five	*POTPOURRI*

Pattern Category One includes the sightings of a scene with several objects or characters engaged in an activity. Categories Two and Three see the pattern collectively as did Homer, as a picture of a single object: *Pattern Category Two* includes the sightings of an animal, commonly a bear, while *Pattern Category Three* includes the instances of an implement—commonly a vehicle or a tool. *Pattern Category Four* makes little or no use of the stars as a picture, but instead musters them more or less in a roll-call of individual objects or characters. And *Pattern Category Five* embraces the sightings that, like the answers in the Potpourri column on the television game show *Jeopardy!*, simply do not fit within the other categories.

To be sure, each of these categories contains plenty of stories that are culturally connected. But not always. As presented here, they collect similar patterns. Any grouping of this type is slippery: a roll-call roster might at some point have been a visible object, and a sighting of one object at one time might have been a sighting of two or more objects at other times. My amateur's organization is neither historical, linguistic, nor anthropological—though ingenious efforts of true scholars in many of the social and physical sciences are credited in the Bibliography, each of them groaning under the weight of its *own* bibliography. Instead, I am making gatherings of similar outputs of the human imagination, which people quite near to and far from one another have applied to this natural appearance.

First, here is an example. In Finland, the asterism has had many names, but one of them is a net. The Numic family of Native American cultures, half a world away, also sometimes referred to the pattern as a net. Look at the seven stars with fresh eyes, and you too can make out a squarish net or weir cast at the end of a thrown rope.

I am not claiming that the Finns and the Numics are genetically related to one another, or encountered one another in history, and thereby transmitted or received the story. Neither traded with or conquered the other. Rather, my point is that humans love to make patterns, and they love to make those patterns based on the objects and ideas that are most important to them. So it is not surprising, and indeed it is to be savored, that two different cultures would each look skyward and think about fishing and hunting.

Second, here is a preview. The fainter Alcor sits amid seven more prominent stars. In the tales, it is variously a dog, a pot, a woman, a girl, a boy, a wagon-driver, a future ruler, and the middle toe of a giant. In Greek it has been the abducted Pleiades Elektra; in Latin it has been Eques Stelluria, the Little Starry Horseman; in English it has been Halopex the Fox, the Rider on the Horse, and Jack on the Middle Horse; in German it has been Hinde the Farmhand and Hans the Thumbkin; and in Hungarian it has been Göncöl the exiled shaman! Across the cultures and across time, Alcor takes an exceptional role in the sagas. It has served as a prize, an object of affection, a fitting climax, a punchline, and the seed of another tale. What a gift a diminutive object is to the weaver of stories.

This approach supports my larger point about how we think about our species. We spend large amounts of time and trouble breaking ourselves down—between regions, countries, classes, tongues, civilizations, bloodlines, genders, and races. It is time to take a break from tribes. Instead, for one evening, let us celebrate all the patterns and stories that humans have seen and told the world over, and relish the fact that different peoples have seen similar images.

Metaphors of a melting pot and reflections upon the essential unity of mankind are out of step today with both populist nationalism and identity politics. They border on the corny and the obtuse. I maintain that a collective view is appropriate especially on this topic, and in mediating between us here on Earth and the stars. When viewed from any appreciable distance, as with the 1968 Apollo 8 "Earthrise" photograph that graced the cover of *The Whole Earth Catalog (see Figure 7)*, our marble is very much one fragile point of faint multicolored light.

Figure 7. The Constellation That Really Counts

IS THERE A GREATER BEAR?

Before we proceed further, I am eager to cast into pale doubt the conceit that everyone sees, or should see, a Greater Bear across a broader Ursa Major constellation. I certainly cannot.

There is no question that a Bear Culture is prevalent across the ancient history of much of Eurasia and North America. However, many of the societies surveyed here do not tell that there really is a large beast up there, traceable in dozens of stars. Some peoples refer to a Bear but they locate that animal only in the 7+1 asterism or a subset, and not in a greater constellation to which the 7+1 merely form the rear end.

The Ursa Major stars other than those in the Dipper are rather faint. Considerable imagination and license are needed to turn them into anything. In some ways, it is a measure of how powerful the cults on Earth have been that so many peoples have wedged a bear into space without any obviously appropriate pictures.

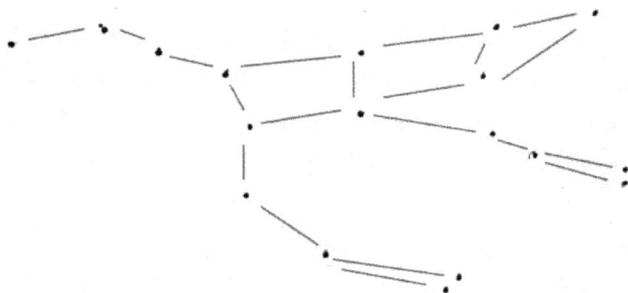

Figure 8. Ursa Major, perhaps

Many candid commentators, from ancient authors to Madeleine L'Engle, have frankly observed that there is nothing in the sky that remotely looks like such an animal. We can picture dippers, wagons, and ploughs, but not so easily a bobtailed bear.

The older Greek writers—Homer, Hesiod and Thales—do not appear to have told tales about such a larger constellation. "Ursa Major" is not their bear. It was the later Greek and Roman writers, integrating myths with their astronomy, who made out the larger setting.

Figure 8 provides one traditional way of connecting the Ursa Major dots to make such a Greater Bear. But many other cultures have had no use for such a large-scale pattern.

Further support for setting the greater Ursa Major constellation aside for now is that no two artists or traditions seem to agree on how to connect the dots to make such a larger animal. *Figure 9* and *Figure 10* show two more ways the picture has been oriented to make a bear with a Big Dipper tail, each heading in a different direction. Look online and you will see several other, even less plausible overlays.

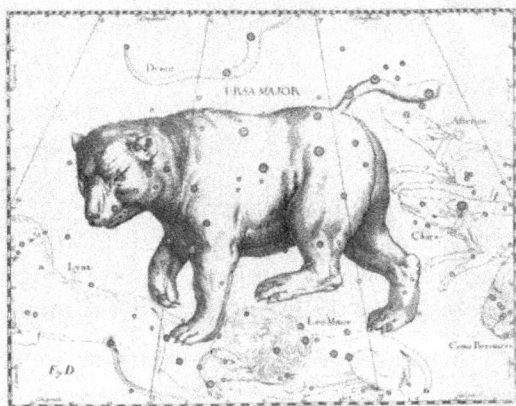

Figure 9. Bear Down

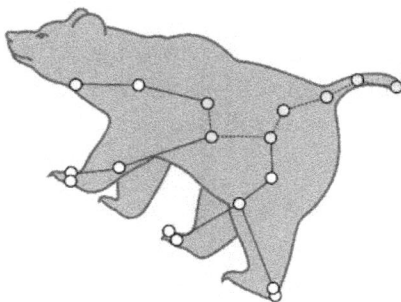

Figure 10. Bear Up

The children's book author H. A. Rey, of *Curious George* fame, proposed a backwards version that has been adopted by many, including the makers of apps for smartphones. *Figure 11* depicts a long-necked polar bear facing the opposite way, with the Dipper acting as a saddle, and the head where the ancient artists have the tail!

We will come to some complex star patterns, including but not limited to that Greater Bear of the later Greeks and Romans. But we need not harbor primal shame for not being able to see a larger beast. As an amateur, I am at peace giving the 7+1 the title and dignity of their own "constellation," and not just those of a subordinate "asterism." For now, it is best to clear one's mind of a greater animal; to un-imagine such a pattern (if indeed you ever truly imagined it); and to see patterns and hear stories based only on the 7+1.

Figure 11. Polar Bear, Backwards, With Same Stars

PATTERN CATEGORY 1: *The Scene*

Case 1A: The Cosmic Hunt

An asterism of a bear is universally recognized by scholars of our earliest human cultures. It is certainly one of the oldest attributions that can be found, as the specialists conclude. Indeed, this may be the oldest pattern in nature that humans ever conceived.

But the bear to which these sources refer is usually *not* the modern astronomer's larger Ursa Major of dozens of stars, the one we have seen in the constellation books. Rather, it is a complete bear either consisting of the seven stars—or instead a subset of those seven!

I was astounded to learn that the principal conception may have been of the four bowl stars as the feet of the bear, followed by the three handle stars. It is this pattern *(see Figure 12)*—a quadruped tracked by three individuals, not the seven-starred Dipper pattern, let alone anything larger—that may be the true Bear relic of the last great Ice Age, peaking some 22,000 years ago. That 4+1+1+1 pattern is regarded by several researchers as the primal scene of our far remote ancestors.

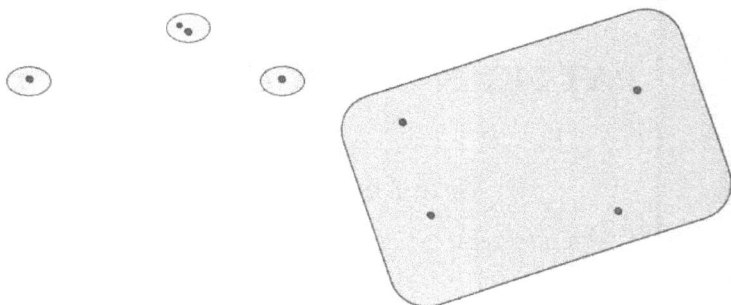

Figure 12. The Original Cosmic Hunt: 4+1+1+1?

Once you see the stars in this configuration, it is impossible to un-see them in this way. We can imagine the boys and girls of countless millennia past, gazing far overhead at this pattern while listening with rapt attention to the tales of their elders.

Although we believe some of the ancients had a specific Bear story, it is a matter of contention which stars were used to play which parts. To add to the confusion, there is evidence that some cultures told the Bear story not about the Dipper but another constellation, like Orion or Boötes. Eventually many of the stories emanating from Eurasia, and Siberia in particular, settled on our Dipper as the asterism where the action takes place, and that is how I have collected them here.

Cave-bear skulls found in Paleolithic (Old Stone Age) mounds more than 30,000 years of age suggest the basis for stories about a bear cult. Cave paintings at Lascaux, France and elsewhere may also depict Taurus the Bull or other animals in the evening sky. Much attention has been paid to an amulet made from a fossilized sea urchin found in northern Europe. The amulet is adorned with careful carvings suggesting the Dipper with the stars in an orientation that is to us slightly out of key, as they might have been aligned long ago *(see Figure 13)*.

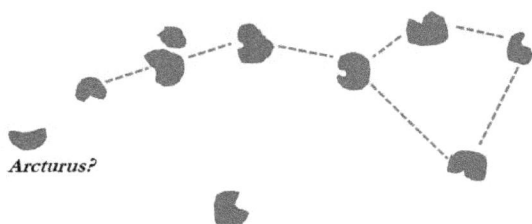

Figure 13. Stone Age Carvings on Sea-Urchin Amulet
(and Possible Connecting Lines)

The Bear is unlikely to be an independent invention by different cultures. This is so for no reason more persuasive than that the four stars of the bowl simply do not look much like an animal either! Some scholars of the Bible, in search of historical evidence for its literal contents, attribute the prevalence of the story to the dispersal of peoples after abandoning construction of the Tower of Babel.

It is possible that the Bear stories worldwide are essentially one story. That tale might have started with tribes that spread across Eurasia over the early millennia.

As they headed northward and eastward into Siberia, the animal in question may have been transformed into the Elk, the major animal native to the northern parts of the continent. The story may also have been carried some 14,000 years ago across the Bering Sea land bridge to the Americas, or more recently by coastal ocean voyages. One scholar muses that these long journeys would have been enlivened by a "chain of grandfathers" telling a chain of grandchildren a tale of a hunt in the sky.

What is the background of this Cosmic Hunt? It is best explained by telling it. I will set forth an extended common tradition, found in eastern North America—told by some members of the Algonquian and Iroquois language groups. The global tale involves a bear—or an elk or other game— followed by several individuals. Often the Dipper's three "handle stars," the ones the astronomers call Epsilon, Zeta and Eta, represent three hunters, or one hunter with two dogs. Little Alcor is regularly included in the form of a diminutive accessory to the hunters, as a dog belonging to one of them or a pot for cooking the game.

One of the most elaborately preserved of these Cosmic Hunt stories is one recorded in 1900 by Stansbury Hagar from two members of the Mi'kmaq tribe of Nova Scotia and environs, associated with the Algonquian language family.[2] To appreciate this story, one must have a proper

---

[2] Hagar was a law student who transcribed stories from two family members in Digby, Nova Scotia. He has been faulted for the small and correlated sample and informal method. Skeptics of a unitary Eurasia/North America bear-hunt story ask where

understanding of the rotation of the newly risen Dipper through the seasons. The one in *Figure 3* that I originally downloaded from the web was incorrect. The correct orientation of the Dipper at its first rising across the course of a year is shown in *Figure 14*. Over the course of twenty-four hours, the Dipper begins at this rising position in the early evening, rotates in apparent motion to us around Polaris, and by the following evening returns close to, but slightly advanced from, that starting position.

---

are the rest of the 14,000 year old sagas, if this one has miraculously been carried down for millennia.

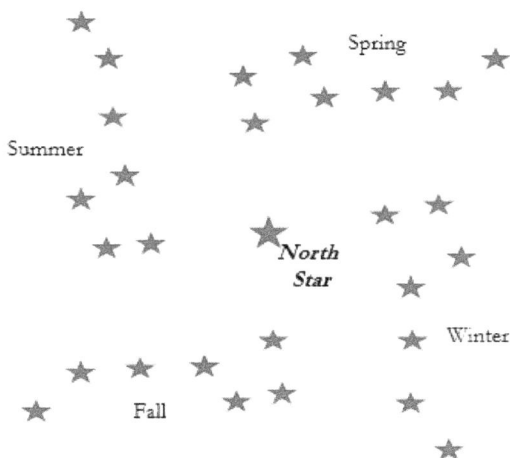

Figure 14. The Correct *Rotation of the Dipper*

I now use the conflicting pictures of *Figure 3* and *Figure 14* in a different kind of teachable moment to instruct my young audience:

Don't trust the Internet.

This tale calls for eight characters:

- *a Bear, naturally, being the four-star bowl of the Dipper;*

- *three nearby bird hunters, represented by the stars in the handle of the Dipper, hot on the Bear's trail so to speak (Epsilon, Zeta and Eta of the handle being Robin, Chickadee, and Moose-Bird respectively); and*

- *four more distant bird hunters, represented by stars in the constellation Boötes, which as you may recall is non-circumpolar—those stars set below the horizon for portions of the year in North America.*

In the spring, the four-footed Bear emerges from the crown-shaped den of Corona, rises at the very top of the sky, and plods nightly around Polaris. Little Chickadee (the middle star in the Dipper handle) spots the Bear and calls for the larger Robin and Moose-Bird (the other handle stars) to flank it as it travels. Then they summon the four Boötes hunters to join them.

All seven hunters steadfastly pursue the Bear through the summer, when it stands on her hind legs. By late summer, though, Boötes sets beneath the horizon at these latitudes, and its four hunters drop out. In contrast, the three handle stars in the Dipper itself are persistent—they never cease hunting.

In the fall, close to the horizon, Robin shoots an arrow into the Bear, splashing bear-blood both back onto the hunter (whence the male Robin's red breast) and onto the nearby forest's autumn foliage. Chickadee arrives to cook the Bear—and his diminutive companion, Alcor, serves as the cooking pot. In some versions of the story, cooking the Bear's marrow-bone causes fat to be exuded, conferring yellow colors on the dying leaves.

In the winter, the Bear lies dead or dormant. It is supplanted by a new or rejuvenated Bear, and new or rejuvenated hunters, with the coming of the spring.

The cosmic hunt is a remarkable story, one that can hold listeners' attention for an entire cycle of seasons. It places the terrestrial hunt for the bear in spiritual congress with the pageant of the skies, and connects the mystery of the animal's and the hunters' resurrections with the mystery of the cycle of the calendar. The astronomer E. C. Krupp remarked that the celestial bear recapitulates our own experience: after all, a bear undergoes a hibernation that is akin to the deaths on our world, and it is stirred back to life in our spring, just as all the renewed youth and new young of the North are following suit.

Case 1B: The Casket and Three Mourners

Other cultures have also pictured a quadrangle and three trailing individuals, but in an entirely different story than that of the Bear or a cosmic hunt. The second case in this pattern category is that of a dead body trailed by three individuals.

Ancient Arabian and Hebrew traditions each refer to three mourners trailing such a body. In Hebrew the name of one constellation or star is *'Ash* or *'Ayish*. This name appears in the Scriptures as celestial evidence of the ineffable and overwhelming power of the Lord—in Job 9:9

("Which maketh *'Ash* (or *'Ayish*), Orion, the Pleiades and the chambers of the south") and Job 38:32 ("Canst thou guide *'Ayish* with his sons?").

As with the Cosmic Hunt, the name has been attributed to more than one constellation or star. Scholars differ on which was originally intended; candidates include Boötes, the Hyades, Venus the Evening Star, and Leo. In the Vulgate translation of the Bible, St. Jerome chose to translate the term as Arcturus in Boötes, and the King James version did the same. However, in later scholarship, *'Ash* or *'Ayish* has been rendered as the Bier, accordingly referring in several Semitic cultures to what Americans call the Big Dipper.

In all fairness, almost any group of stars that appear to enclose a polygon could represent either an animal or a body. In the desert night, accompanied perhaps by wine, one could see anything anywhere. But these two groups of stories, the Cosmic Hunt and the funeral procession, have both consolidated around our seven stars.

A common translation of the Arabian name *Banat Na'ash al Kubra* for the handle stars collectively is the Daughters of the Bier, with the elder daughter *Al Ka'id,* the Chief One, the star we call Alkaid. The seventeenth-century astronomer Johann Bayer adopted *El Keid* as the name of the entire constellation. Certain Bedouins are said to speak of the two Dippers as two biers. In one Arabian treatment of the funeral story, the diminutive Alcor was a newborn infant daughter of one of the mourners.

The funerary tradition continued into medieval times. Arabian Christians in later centuries are reported to see the Bier of Lazarus, illustrating the New Testament miracle recounted in John 11:1-46. In Europe, Italians also are said to call it *Cataletto*, the Bier.

Different funeral stories arose in other places and times. The Koryaks of far eastern Siberia are reported to have referred to an *arangas,* a rack on stakes used only for especially respected corpses. The rack was trailed in the sky by three mourners, who are the three widows of a dead divine chieftain. In this tradition, the diminutive Alcor is a child who is the ancestor of the chief of the terrestrial tribe. Thus the smallest in the heavens today is to emerge as the greatest on Earth, in the fullness of time.

The funeral pattern and story show up in North America as well. The Pawnees have referred to stretchers; the Osage, once again to a bier. A tradition within the Sioux nation refers to a funeral bier and three mourners. In some traditions, they are joined by Alcor the smallest sister, said to be crying.

Case 1C: Other Accompanied Quadrangles

There are stories interpreting the 4+1+1+1 pattern as other kinds of quadrangles. Separate from the Cosmic Hunt, there are traditions where the quadrangle is an object and the tail stars are individuals. Native American groups in Quebec were said to regard the asterism as a Canoe followed by three individuals. In the inland plateau of the Pacific Northwest, a Coeur d'Alene tribe is reported to tell of a grizzly bear—and its three brothers-in-law.

An obscure tale of the Votyak and Udmort peoples from the Volga delta region of Russia makes use of a sky table with three legs. A Tunguskic tribe in far eastern Siberia saw in the asterism a bed and three family members. And there is a Bengal or Hindustani tradition referring to a bed and three thieves.

* * *

Once you have seen the pattern as 4+1+1+1 and heard the saga of the Cosmic Hunt, it is indelibly coequal with the seven that we have been taught to detect. It is the 7(+1) patterns to which we now turn, in *Pattern Category Two* and *Pattern Category Three*.

PATTERN CATEGORY 2: *The Animal*

Case 2A: The Greater Bear (Ursa Major)

The larger constellation Ursa Major we have seen in the astronomy books, traced across of a dozen stars, makes its debut in renditions of the Greek myths that were first compiled well after the time of Homer. The Greater Bear name is so pervasive today, however, that it is time to tell the tale of that extended figure.

Let us begin with a little linguistics and Greek literary history. The Indo-European family of languages stretches from Ireland to India, from Iceland to Iran. One common feature of much of this landscape, and in particular of the suspected homeland of the Indo-Europeans, is the bear. The Proto-Indo-European (PIE) linguistic root for "bear" is believed to be **rkto-*. It shows up in Sanskrit as *riksha* or *rkshas* and in Greek as *arktos,* as in the epigram from Homer quoted above.

Arktos is also the Greek word for "north," and thus the immediate root of our English word "Arctic." But the character *Arktos* is also the Greek mythical figure who was guardian of the bear. And *Arcas* is said to be the progenitor of the Arcadians, who are variously said to be the people of the bear, the people of the hills, or the people of the north.

Was north named for the bear or was the bear named for north, and were the people named for either? The linguists say that "bear" came first, though the speculations of the Greeks at the time appear thoroughly commingled. Aristotle gives one name of the north wind as *Aparctias,* from the land of the Bear. He remarks that the regions below the Bear are too cold to live in. Strabo drew a similar connection from the fact that the bear is the only large creature that dares to inhabit the frozen North.

Homer seems to have considered the seven stars alone to constitute *Arktos*, the Bear, since he says that same pattern is "also" called *Hamaxa,* the Wain. That was the prevailing Mesopotamian term exclusively for the seven-star pattern. Homer mentions Orion, the Pleiades, the Hyades, Sirius, and Arcturus—and that it is it for his starry roll-call. No longer roster is found in the early writings attributed to Hesiod (the *Astronomia* and the *Eoxae*, his "Catalog of Women") or in the Homeric Hymns. That list is the entire extent of the recorded northern pattern names in the 1000-800 BCE period of ancient Greece.

The polymath Thales of Miletus (circa 600 BCE) is reported to have filled the sky out with some more constellations, imported from Mesopotamia, the Levant, and perhaps Egypt. Ancient sources say sailors brought him the Phoenician Wagon or *Phoenike,* also known as *Cynosura* and *Ursa Phoenike;* this is the bear closer to the North Star, and thus the one more useful for navigation. Only later did this constellation become known as Ursa Minor. At some point, the original bear was given the name Ursa Major as a counterpoint.

By the time of Eudoxus of Cnihus (390-337 BCE) and his *Phainomena,* a poem now lost, there were names for both constellations. His roster included the Greater Bear and Lesser Bear groups themselves, each extending beyond its original seven stars.

Although Eudoxus's work has disappeared, other Greeks made use of his system and roster. Aratus of Soli (315-240 BCE) wrote the poem *Phaenomena* with both extended Bears. Works attributed to Eratosthenes of Kyrene (275-194 BCE) (hence sometimes called "Pseudo-Eratosthenes") and Hipparchus (190-120 BCE) completed the Hellenistic sky picture in which these two animals spanned much of the heavens.

In the Imperial period, we have seen that Ptolemy (100-170 CE) wrote *Almagest;* that work refers to *Arktos Megale,* the Greater Bear, and a Lesser Bear as well. The Romans liked and continued the Greek myths. The works of Hyginus, Manilius, Germanicus Caesar, Strabo, Ovid, and even Cicero carried forward the system of Aratus.

Enough of the intimidating names of distinguished authors—what are the classical tales associated with the Greater and Lesser Bears? I was not surprised to see quite a few variations. Each source confidently tells one and only one version, while rarely acknowledging that the others exist.

The most famous Greek stories had been passed along to Rome by the time of Ovid (43 BCE-8 CE), whose *Metamorphoses* and *Fasti* (a work about the calendar) record or more likely improve upon them. These stories congealed around a young woman, in most cases named Kallisto, daughter of Lycaon, King of Arcadia in Pharsalia in the central Peloponnese region. In many tellings, her virginity was sanctified to the goddess Artemis, goddess of the hunt.

The woman's name may stem merely from an epithet: *Kalliste Artemis* would mean "the most beautiful Artemis," and may have initially been simply an appellation for the goddess herself. Later, a Kallisto may have been personified. In Ovid's telling for the Romans, Zeus, Hera and Artemis of course become Jupiter, Juno and Diana.

This Kallisto is said to have been lusted after and ravished by Zeus, who took on the appearance of Artemis herself until the last moment, as it were. When the "horns of the Moon were … rising again in their ninth course," Artemis was bathing with her consorts. The goddess beheld Kallisto's expectant state for the first time, and promptly banished her. (Kallisto's ravishing, and the revelation of her pregnancy, were the subject of famous oil paintings by Titian, Rembrandt, Rubens and Boucher, depicting arresting scenes from classical literature.) Sequestered from the other females, Kallisto gave birth to her son, Arcas.

The sky-god's furious and often-wronged wife Hera continued to pursue and harry Kallisto. She was turned into a terrestrial bear. But by whom? The accounts vary— by Hera (as in Ovid), or by the *real* Artemis (as in Hesiod, Hyginus, and Manilius), or by Zeus himself to shield her from both Hera and Artemis and from an even worse fate (as in Eratosthenes).

Traditions hold that many autumns later her human son Arcas, now grown to manhood and occupied as a hunter, espied his ursine mother. Kallisto reared up on her hind legs to greet him. He was about to shoot or spear her, or hit her in the head with his shield. Many tellings report that before he did so, Zeus cast her into the skies. The vengeful and persistent Hera induced Oceanus and Tethys, Kallisto's foster parents, to agree never to let their poor daughter cool her paws by wetting them in the sea.

Arcas has many different fates in these stories. In a few, he is immediately thrown among the stars along with Kallisto. More typically, he lives to maturity in order to participate in other myths. Later he is placed in the heavens forever to be with his mother, ambiguously in Ovid as *vicinaque sidera* or "neighboring constellations."

But *which* neighbor is Arcas? Most have interpreted Ovid to refer *not* to the Lesser Bear but instead to Boötes. According to Matthew Arnold's version, they are "the guard Arcturus [and] the guard-watch'd Bear." So that tale

I told the kids about the Big Bear and the Little Bear being mother and son is incorrect from the standpoint of many renderings.

An immediate objection to these 7+1 bear stories is that the pattern of each Dipper clearly has an elongated "tail," and our terrestrial bear obviously clearly does not. Even my scouts must have been a little dubious when I described the tails.

Something cries out to be done about this discrepancy. With characteristic resourcefulness, the sources suggest that Zeus lengthened the tails to their extraordinary sizes in the course of grabbing the bears and flinging them skyward. An English author and Fellow of Trinity College, Thomas Hood, pragmatically suggested in 1590 that "there was great likelihood that [in throwing Kallisto into space,] her taile must stretch. Other reason know I none." Well, know we none either.

The two bears together have also been seen as forming a kind of super-constellation, a wall separating parts of the sky. Vergil, in the *Aeneid*, speaks of twin oxcarts, *geminosque Triones*. But in another poem, in the *Georgics,* Vergil reverts to the Ursas Major and Minor: "around our poles the spiry dragon [Draco] glides, and like a wand'ring stream, the Bears divides."

As with Vergil, so unsurprisingly with his disciple Signor Alighieri. As he passes into heaven, Dante's narrator in *Paradiso* records that "Minerva breathes, and pilots me Apollo, and the Muses nine point out to me the Bears."

Case 2B: Other Bears

Other cultures have perceived just the seven-light pattern, with or without a larger conglomeration, as a bear. In Turkey, the stars have formed the Great Bear, perhaps from the same sources as the Greek tradition. There are Hebrew and Arabian traditions calling the pattern the Bear (*Dobh*), competing with the Casket stories we saw above. The Turuchunsk of Central Asia are said to have referred to a hunt and a bear, this time with Polaris being its hunter. Some Finns call both the constellation and the bear *Otava*. But is it the bear that names the stars? Or is it the stars that name the bear?

The Bear also appears throughout Romance language and folklore: in Italy as *Orsa Maggiore*, and in France as *Grande Ourse*. Federico García Lorca's poem "Canción para la luna" refers to *Osa Mayor*. In the sixteenth century, Luis de Camões makes use in *The Lusiads* of both the Bear tradition and the Chariot tradition I describe below. Here is a rather old-fashioned English translation of the Portuguese original:

Lo, bright emerging o'er the polar tides,
In shining frost the Northern Chariot rides;
Mid treasur'd snows here gleams the grisly Bear,
And icy flakes incrust his shaggy hair.

The Bear pattern is recognized broadly but infrequently in northern Europe, including Germany (*Großer Bär*), Scandinavia (*Stora Björn*) and the Netherlands

(*Grote Beer*). It appears that the Implement names described below are more common in those countries.[3]

In Sanskrit, the words get complicated. *Ṛikṣa* as a feminine noun has a meaning "bear," but as a masculine noun has a distinct meaning "star" or "shine." Early on, there may have been some mixture of the two meanings. Moreover, *rikṣa* is close to *riṣi* or "sages," and thus to the "seven sages" used in a roll-call pattern discussed under *Pattern Category Four* below.

Many tribes of the indigenous peoples of North America refer to a Bear. Iroquois members surprised French missionaries like Chrestien Le Clerq (1691) by saying that "the French bear" is also "our bear"—called *okouari*. The Native American Bear was separately reported by Roger Williams (1643) (among the Most), Cotton Mather (1712) (among the Paukonnawaw), and Silas Rand (1850) (among the Muen). Far in the west, the Southern Paiute, the Zuni, and the Keresan are all reported to call the pattern a White Bear.

It is time to check in on the appearances in English literature. Shakespeare mentions the Bear in *Othello*:

---

[3] Note also that many European languages came up with respectful euphemisms for this intimidating animal, such as "the brown one" (Germanic *Bruin* or *Bär*) or "the honey-eater" (Russian *medved*), such that the Proto-Indo-European root does not appear in words in those tongues except in learned borrowings from Greek or Latin. This is something like the non-name name for *you-know-who* in the Harry Potter mythos.

The wind-shak'd surge, with high and monstrous mane,
Seems to cast water on the burning Bear,
And quench the guards of th' ever-fixed pole[.]

We will see that Shakespeare also calls the pattern by several of its other names.

Case 2C: The Elk

Some scholars think the subject of the Hunt myth in Northern regions was not a bear but another prevalent hunted animal, an elk in Europe or a moose in America. The same type of story would be told, namely that when a particular beast was killed, it ascended into the sky.

The Evenks of North Asia are said to have called the constellation *kheglun,* the Elk. Ostyak peoples, near the Yenisei River in central Siberia, are said to have spoken of *Vitim-Olekma,* a pregnant six-legged elk pursued in the sky by a hunter whose snowshoe track was the Milky Way galaxy. When the extraterrestrial elk leapt to Earth, the hunter managed to lop off two of her legs, thus explaining the difference between the earthly and heavenly varieties.

The Lapps in northern Scandinavia are also reported to have called the pattern the Elk. There is a Russian Mason-Dixon line of sorts between the Animal stories and the Implement stories: the northern steppes have tribes that call the pattern *Los*, the Elk, while some tribes in the southern steppes call it *Kola*, the Cart.

Case 2D: Other Mammals

Bears and elk are not the only mammals that humans have pictured in the seven stars. We have already seen that the Arabic names for the stars include references to parts of an Eastern Sheep. Syrians called them the Wild Boar. Other desert tribes have referred to the two Dippers as a gazelle and its young. Berber tribes called each of them a Camel. The Tuareg of the western Sahara knew Ursa Major as a mother camel, in one telling a camel owned by Noah. For some, the two Dippers are a mother camel and its young, tethered or led round Polaris.

Lapps at one time called the constellation *sarv*, their culture's all-important Reindeer, while Inuit in North America named it the Caribou, *tukturjuk*. Yakut Koryaks of far east Siberia called it both the Reindeer and the White Bear. Other Siberians referred to it as a Stag. Some Estonians called it the Ox. Minoans probably called it an ox (later a wagon), as in any event there were zero bears on Crete.

In America, a population of Paiutes in the Great Basin called the pattern the Mountain Sheep. The Hidatsa of North Dakota called it the Ermine. A Sioux group are said to have called it the Skunk. Certain Mexican cultures are reported to have seen the pattern as an ocelot. Ojibwas called it the fisher or marten, with Alcor as an irritating arrow in that sleek animal's tail; this reference appears in the source materials for Henry Wadsworth Longfellow's poem *Hiawatha*.

The Turkic Teleuts in south Russia had an entirely different way of looking for an animal in the sky. They referred not to a physically present animal, but instead to tracks. The quadrangle corners represent the four hoofprints of a horse, and the three stars the *swish-swish-swish* sweep where its tail had brushed against the darkness.

Case 2E: Other Animals

Finally, a menagerie of animals other than mammals has been imagined in this grouping.

An early Egyptian pattern was not a complete animal, but just the thigh and leg of a bull whose name was Meskhetiu. The constellation was invoked, along with an adze that itself was in the shape of the Dipper, by the god Set in connection with the death and rebirth of Osiris. Eternal life was naturally associated with the circumpolar motion of these stars. Later Egyptian traditions called the asterism the Dog of the god Set, or part of the Cart of the god Osiris.

The Mayans in the mythology recorded in the Popol Vuh text called the pattern a bird, the Seven Macaw (*Waqub' Kaqix*). The bird served as eternal antagonist to their hero Junjapu, corresponding to the planet Venus.

One group of Native Americans was said to have called the asterism a Brood Hen, which is difficult to picture. Some South Americans, when they could see the constellation, were said to envision their own resident large animal—a big landbound bird, the Rhea. A seafaring group in the East Indies referred to it as the Shark. Like the Finnish Net and the Numic Net, the name given to leading stars will often be an animal or object that is key to a culture's existence.

Aztecs called the 7+1 pattern the Scorpion. Incidentally, another Aztec tradition depicted the seven stars as the warrior night god Tezcatlipoca, but with an exposed bone where a left foot should be. At the latitude of Mexico, one star of the Big Dipper tail indeed sets below the horizon, hence the missing limb.

Burmese peoples called the pattern *Pucwan Tara,* a type of crustacean, like a shrimp, crab, or lobster. Since Burma is low in the northern hemisphere, the asterism there is not circumpolar—it rises and sets at fixed times each year. This is an advantage not a disadvantage for farmers, because the constellation's rising on the horizon at a particular time was useful for calculating when to plant and harvest crops.

Thus, I was wrong in teaching my youthful charges that being circumpolar is always preferable. Weather-watchers would rather learn about a constellation that is not circumpolar. For example, they witness the first reappearance in northern latitudes of Sirius and the Greater Dog, Canis Major, in the month of July. That star-rise launches what we still refer to as the dog days of summer.

PATTERN CATEGORY 3:
The Implement

It is likely that some of the Scene and Animal stories about this star pattern, those in *Pattern Category One* and *Pattern Category Two,* were the first celestial pictures to be formed. Advanced tools came long after we had exposure to animals.

Farming began in earnest around 10,000 BCE and the wheel was invented around 4000 BCE; the axled wheel, a bit later. Chariots, wains, wagons, carts, and ploughs emerged shortly after. The celestial Implements originated around this time as well.

Case 3A: The Wagon

Already in Homer, we know that the asterism is "called by some the Wain." The Wain, *hamaxa* in Homer's Greek, is the equivalent to the Akkadian term *Mar.gil.la,* used at roughly the same time in Mesopotamia. A Middle Eastern inscription dated to 1700 BCE resembles a Wagon. Babylonian and Assyrian descriptions of a Wagon pattern are present in cuneiform tablets and other texts.

The Latin term for each of the Dippers is *septentriones*, *septem-* + *-triones,* which means "seven plough-oxen." We have already seen Vergil refer to the twin ox-ploughs, *geminosque Triones.* The septentrion term came to be known as just the number seven, thus covered in the Roll-Call category below. But sometimes the plough-oxen were represented by a scene instead, each of a cart being pulled by two oxen—one cart for each of the two Dippers.

Romans also called each of the seven-star Dipper patterns the *plaustrum*, the Wagon. This Wagon reference continues across Europe—in Romance languages, like the French *le chariot grande,* the Portuguese *la carreta,* and the Italian *il grande carro.*

In German, the Big Dipper is *Große Wagen*; in Swedish and Icelandic, *Stori Vagn;* in Estonian, it is *Suur Vanker.* In Romanian and other Eastern European traditions, it is some other variant upon the Great Wagon. Sometimes it is more specifically the "wagon of heaven," in some cases connected with Elijah's Chariot of Fire mentioned in the Bible, illustrating the spectacular scene in II Kings 2:11. Thus we have the *Himmelwagon* in German and the *Woz Niebeski* in Polish.

Further afield, we find a Hungarian reference to *Göncöl's Wagon* or *Big Göncöl,* after a shaman (*táltos*) who possessed medicine that could cure any disease. Alcor is of course the shaman in question. And in China, the stars could be made out as the Chariot of the Emperor of Heaven.

In early Irish sources, the asterism was labeled as King David's Chariot. There is also a 1593 reference to *Camcheacta* in Gaelic. In Saxon regions, where Irmine was a war god like Ares or Mars, the pattern was known as *Irmines Wagn*.

It is time to turn more directly towards our English language names. The term *Wain* is ancient, dating according to the *Oxford English Dictionary* (OED) to 888 and the translation of Boethius attributed to King Ælfred the Great ("þe we hatað wænes ðisla") and then to Chaucer's use of the term around 1387 ("woot nat whi þe sterre boetes passeþ or gaderiþ his weynes"). The earliest reference in the OED to *Wagon* with a starry meaning, by contrast, is not until 1867.

Across northern Europe, the two wagons of the Big Dipper and Little Dipper have been more specifically possessed. The smaller asterism has been feminized in some northern languages as *Kvinnowagen* or *Kevennavagn,* meaning Woman's Wain. Deducing the origin of that name is comparatively easy.

The Big Dipper, on the other hand, has been more variously described. It has been called a wagon or wain, but belonging to whom? To a man, to a king, or to a god?

In Norse languages, the asterism has been called the Wain of Odin or Woden. Since the time of the Franks, the constellation has been known in northern Europe as

Carlswæn, Carleswaen, Karl Wagen, or *Karlavagn.* All of these can translate to *Charles's Wain* or just *Charles Wain,* and that term was in fact used for centuries in England and in colonial America.

The label may have begun as a reference to the deity named *Karl,* an alternative name for Thor. It naturally came to refer to the Frankish Emperor Charlemagne, or Carolus Magnus. Indeed, the OED cites an appearance of *Cherlemaynes Wayne* in 1398. A folk etymology suggesting the term was originally *Churl's Wain* (i.e., Man's Wain) has been hotly disputed.[4]

The early English term *Arcturus Wain* gave ownership of the wagon to the guardian, the main star in Boötes. In early Britain, that name might have been connected to Camelot as *Arthur's Wain.* The Normans might have brought *Charlemagne's Wain* from the continent. *Arthur's Wain* was in any event displaced as the standard English literary reference by the fifteenth century. From that time, the OED indicates that the Ursa Major asterism has been most consistently referred to as *Charles's Wain* or the equivalent.

Literary references in English to the Wain or Wagon abound. The Coverdale Bible (1535) refers to "the Waynes of Heaven." In Walter Scott's intentionally anachronistic *Lay of the Last Minstrel* (1805), this old name is used:

---

[4] Both Webster's 1890 dictionary and Wikipedia claim *Charles's Wain* originates from *Churl's Wagon,* "Man's Wagon," citing Swedish sources. The OED rejects the etymology proposed by these texts, at least for English usage.

"Arthur's slow wain his course doth roll,/In utter darkness, round the pole." King James I of England wrote poetry referring to "charlewain." Courtiers of the House of Stuart clarified that the constellation should be *King* Charles's Wain (now alluding to King Charles I himself, guiding his country as England's Wagoner). The Wain or Wagon continues to appear, with the last reference in the OED to *King Charles's Wain* being 1876.

Edmund Spenser's *Faerie Queene* (1596) appropriated Homer's image of circumpolarity:

By this the Northerne wagoner had set
His sevenfold teme behind the stedfast starre,
That was in Ocean waves yet never wet[.]

Shakespeare in *Henry IV, Part 1* described midnight as the time when "Charles's wain is over the new chimney." Alexander Pope conferred ownership of the wagon on the cowherd in a couplet: "So fares a sailor on the stormy main,/ When clouds conceal Boötes' golden wain."

Alfred, Lord Tennyson in the 1832 poem "New-Year's Eve" used this name for the pattern in the line "Till Charles's Wain came out above the tall white chimney-tops." In 1850's "In Memoriam," he wrote that the "lesser wain is twisting round the polar star." But in "The Princess," the poet laureate instead referred in 1847 to the beast: "I paced the terrace, till the Bear had wheel'd/Thro' a great arc his seven slow suns."

In some of these Wagon traditions, the diminutive Alcor is the wagon-driver. It is thus neither an implement, nor a child, nor a woman, but instead the vehicle's master.

Samuel Johnson's landmark 1755 dictionary identifies the constellation as being referred to both as the Bear, citing Thomas Creech's 1700 translation of Manilius, and as Charles's Wain. He defines *Septentrion* only as an adjective meaning northerly. *Cynosure,* on the other hand, is crisply defined as what we call the Little Dipper. The good doctor does not give celestial definitions for any of the terms Chariot, Wagon, Wain, Plough or Dipper.

Case 3B: The Plough

In Britain and Ireland, especially in the countryside, the seven stars at some point became a different kind of farm implement, the Plough. Perhaps this designation dates to the Latin *triones.* Germanicus Caesar translated Aratus and frankly commented that he could see a plough better than he could see a bear. In a love elegy of Propertius, the speaker ardently seeks the romantic favors of Cynthia well into the night, without success though "the oxen of Icarus [another name for Boötes] are now slowly turning the stars to the morning."

It may be that rural (and unlettered) folk called the constellation the *Plough* or the *Starry Plough,* while urban (and literary) folk called it *Charles Wain.* Thomas Fale records use of the term Plough "by countrymen" in 1593. The OED references in literature to a *Plough* date only to 1868.

The Starry Plough flag of the Irish Citizen Army of 1916 pays homage to this rustic figure. So do the emblems of later Irish political organizations and many signboards over inns and public houses, including the Starry Plough in Berkeley, California.

The term does not seem to have wide circulation in the United States. American dictionaries like *Merriam-Webster's New Collegiate* feature astronomical definitions for Dipper, Big Dipper, Charles Wain, Wain, and even Wagoner—but not for "Plough."

Other farm implements have been seen around the globe: yokes, scythes, flails, rakes, mowers, and sickles. In India, the constellation has been referred to sometimes as a Plough and sometimes as a Yoke. Even fictional languages get into the act. J. R. R. Tolkien was often vague in describing in what sense Middle-Earth is our planet. But in his "Morgoth" essay in *Myths Transformed,* he hailed Valacirca the Sickle of the Gods, an ancient elvish name for what Tolkien himself called the Plough.

Case 3C: The Household Utensil: Dippers and Others

At long last, in Case 3C of my system, we come to the grouping in which the Big Dipper label itself resides. This quintessentially United States name is not a term either of ancient vintage or of widespread further adoption.

The expression is not found in Noah Webster's original 1828 dictionary or, as noted above, in Samuel Johnson's 1755 dictionary. It appears to be a coinage of the early new Republic.

The earliest major dictionary found to contain a citation to the Dipper is the 1890 edition of *Merriam-Webster's International Dictionary* in the United States—not as an entry in its own right, but in the surreptitious form of a side remark. Under the proper entries for both Charles Wain and Wagon, there is a brief note to the effect that the pattern is also "commonly called the Dipper."

The earliest citations of literature in the dictionaries are to the work of American writers of 1840s New England. These include a Lowell, Massachusetts newspaper entry in 1842 ("the Dipper in the Great Bear"), and rather prosaic comments by two nonetheless important authors, both penned in 1848 though published much later:

> "The comet makes a great show these nights. Its tail is at least as long as the whole of the Great Dipper, to whose handle, till within a night or two, it reached in a great curve, and we plainly see stars through it." (Henry David Thoreau)

> "The stars came out, and the constellation of the Dipper hung exactly over the Val d'Arno, pointing to the North Star above the hills on my right." (Nathaniel Hawthorne)

Most dictionary references are to a single and unmodified "Dipper." Perhaps the Ursa Major seven-star pattern was the Dipper first, and later it was renamed the Big Dipper as a retronym, along with abandonment of the name Cynosure and substitution of the term Little Dipper.

More cookware abounds. The principal Netherlands term is the Saucepan (*Steelpannetje*), reportedly more prevalent than the Bear (*Grote Beer*) cited above. In France it has been known as the Ladle or the Casserole. A Lithuanian and Estonian term is the Bucket. An early northern English term is the Butcher's Cleaver.

Nor is the utensil reference limited to the United States and Europe. A recent Chinese term is *Beidou*, the Northern Dipper. In Malay, it has been *Buruj Biduk*, the Dipper.

Case 3D: The Drinking Gourd

We come now to a matter of some delicacy. I use the term "delicacy" with some considerable delicacy myself.

Numerous works in the last seventy years—novels, children's stories, planetarium shows, plays, and musical performances—concern a folk song, "Follow the Drinking Gourd," that was in wide currency in the southern United States prior to the American Civil War. It is said that enslaved people knew the Big Dipper as the Drinking Gourd, and used the asterism to point the way to Polaris, to the northern United States and Canada, and to freedom. More ingeniously, though, the lyrics of the song contain references to a time of year to begin the journey ("when the first quail call") and directions to take ("the river ends between two hills"). In some tellings, the lyric was passed from plantation to plantation by an itinerant blacksmith whose peg-leg left a unique trail to follow as well. The Gourd name, the pointers to Polaris, the song, and the coded lyrics together were celebrated resources of the Underground Railroad, through which many escapees secured their liberty.

It is a very exciting story. Indeed, it sounds almost too good to be true.

Where might the Drinking Gourd name for the seven-star pattern have originated? The *New World Encyclopedia* takes the story as fact and states, without attribution, that the Big Dipper "figuration appears to be derived originally from Africa, where it was sometimes seen as a drinking gourd. In the nineteenth century, runaway slaves would follow the Drinking Gourd to the north and freedom."

What is the evidence for the name "Drinking Gourd"? None is known before the twentieth century. What is the evidence from the nineteenth century for a song about a Gourd being sung? There is none. Beyond that, what is the evidence for any song containing astronomical clues? There is none.

A recent academic work published by a university press cites an impressive-looking source published by an distinguished astronomical society:

"Follow the Drinking Gourd: Bibliography," 89 PROCEEDINGS OF THE ASTRONOMICAL SOCIETY OF THE PACIFIC 291 (1996).

But the only authority in that bibliography for the name, the song, or the story is a children's book—Jeanette Winter's *Follow the Drinking Gourd*, published in 1988. The children's book cites no authority, of course. One's suspicions are further aroused.

An enterprising and careful fellow amateur researcher, Joel Bresler, has published his tentative conclusions on the

Drinking Gourd online (at followthedrinkinggourd.org). He marshals the following facts:

1. Harriet Tubman of Underground Railroad fame was in fact reported to have sung songs with coded messages advising escapees either to move or stay. One song ("a song of peace") would be used if the coast was clear for heading north; another song ("Go down, Moses") would be used if she recommended waiting. But the modern lyrics of the "Follow the Drinking Gourd" song would not have worked as a code—they explicitly mention "the road to freedom" and would thereby spill the beans.

2. The clues supposedly embedded in the currently circulated Drinking Gourd lyrics—a path along a river bank, ending between two hills, leading over another hill to a river crossing—would only have worked for travel from specific places in Alabama to a specific ford across the Ohio River. Underground Railroad escapees typically headed toward cities in any direction, and in any event covered the entire border, not just this one route.

3. The earliest known evidence in print of the pattern name, the song, or the story is the publication in 1928 of music and lyrics to a song titled "Foller De Drinkin' Gou'd." It was published by a white biology professor and amateur folklorist, H. B. Parks, in *Publications of the Texas Folklore Society*.

4. The tale become well known only after 1947, when folk singer Lee Hays told the story, published the song, and sang it in concerts as a member of the Weavers featuring Pete Seeger. The Weavers recorded their version in 1951. Randy Sparks wrote lyrics to another version in 1955, which he played with the New Christy Minstrels and recorded in 1963. John Coltrane, Taj Mahal and perhaps two hundred other recording artists have made further covers of Hays's and Sparks's versions. Adult novels and plays, planetarium shows, and children's books appeared only after the folk song. It appears in over 75 music books. Most of these publications now refer to the song as "Traditional."

5. Bresler found no nineteenth-century sources calling the seven stars a Drinking Gourd, or citing a song with direction instructions. Certainly many people knew to use the seven stars to guide them northward, but we have no evidence of the Gourd name or of an encrypted lyric.

Parks and Hays each claimed to have heard the Drinking Gourd story and song from elderly African-American sources. It is always possible that some earlier individual did write and sing such a coded song. After all, the Underground Railroad did not publish its proceedings and tactics. We cannot prove a negative; absence of evidence is not evidence of absence.

However, there are plenty of postwar histories and memoirs about the Underground Railroad, which mention a large number of operational details, apparently without mentioning such a song. For example, William Still's comprehensive 1871 memoir of the Underground Railroad makes six references to escapees and conductors making use of the "North Star," but not one reference to the Dipper—let alone to a drinking gourd, and let alone to cryptic lyrics.

Even the largest books on star lore published during the nineteenth century and early twentieth century do not mention the gourd name for the pattern, let alone this song or story. Works printed in 1857, 1899, 1911 and 1920 contain no such reference. The closest citation, from 1911, is this more general statement:

> During the Civil War, escaping slaves and Northern prisoners directed their way to a harbor of refuge and home by the friendly beams of Polaris.

Some teachers' plans and planetarium and astronomical education pieces have backed away from the tale. Henry Louis Gates, Jr. of Harvard and Scholastic Books have both discounted the existence of such a coded song in antebellum or Civil War times. Scholars have discouraged reliance on the encryption story, given the state of evidence. A recent teachers' magazine makes a small pivot by asserting that the Big Dipper was known to many as the Drinking Gourd, but offers no proof of how many actually heard a song about that gourd, or of a song with coded lyrics.

The folk musicians of the nineteen-fifties brought the story to new generations of audiences, in concerts and at political assemblies, with lyrics and melodies descended from the 1928 publication and the later recordings. As Bresler says, it is "a song that played a rich role in the folk revival and civil rights movement, and that continues to be widely performed and recorded today."

It is still a wonderful story. With the evidence we have at present, however, it is a twentieth-century story.

.

PATTERN CATEGORY 4:
The Roll-Call

When we enter *Pattern Category Four,* we leave the visual pictures of *Pattern Categories One, Two* and *Three* far behind. We are in the realm of arithmetic and abstraction, where each of the stars stands for an individual character without being part of a visual pattern.

Seven is an especially powerful number in story-telling and myth-making. It is a lucky number, a magic number, in many human endeavors. The Dipper provides a ready-made natural appearance of this digit.

Here we must be careful, because sometimes the stars in the Pleiades cluster are referred to as the "Seven Sisters." Only *six* of the hundred or so cluster-members are readily visible to the naked eye, however.[5] An old Japanese text refers to *mutsuraboshi* or the Six Sisters. Look at the Subaru automotive logo with its six stars; *subaru* is an alternative name for that cluster in Japanese, in that case commemorating the merger of five automotive companies into one.

Meanwhile, the Big Dipper has an *eighth* member in the form of little Alcor. Consider the possibilities for the story-maker created by the two constellations: one pattern includes seven bright stars with a faint eighth companion, while the other has six readily visible stars. Now consider that magic number seven again. The imaginations of story-tellers were stoked, and what followed was a flow of stories of abduction, adoption and marriage.

There have been many public-houses in the west and south of England named the "Seven Stars." The Dipper appears on some of their sign-boards. According to fellow amateur researcher Dr. Hugh Kolb, the original reference for these pubs is more likely the Pleiades—envisioned as a cluster of the luscious grapes of Dionysus or Bacchus.

Here we go. I begin with the Roman terms and proceed to call the rest of the global roll-calls.

---

[5] The text attributed to Eratosthenes explains that only six of the seven Pleiades are visible because the seventh sister, Merope, fell in love with a mortal and is now invisible. I am not sure this counts as much of an explanation.

Case 4A: Septentrion

The principal Roman name for each of the Dipper asterisms is the *septem triones* or *septentrion,* the Seven Plough-Oxen. Even at the time, the term also referred to the compass direction, namely north.

Chaucer translated the Latin *septem triones* into English as the Northernmost constellation in his version of Boethius ("And eke þis Nero goueyrnede by Ceptre alle þe peoples þat ben vndir þe colde sterres þat hy3ten þe seuene triones; þis is to seyn, he gouernede alle þe poeples þat ben vndir þe parties of þe norþe"). The same term was used exclusively as a direction in the "Monk's Tale" in his *Canterbury Tales* ("Bothe Est and West, South, and Septemtrioun").

Dante in his *Purgatorio* used the term *il settentrïon* for the Big Dipper or Little Dipper constellation itself: "the Septentrion, … which never either setting knew or rising, … motionless halted." In some translations the term "septentrion" is fittingly spelled out more completely as "seven cold oxen."

Later, "septentrion" was entirely divorced from any reference to either constellation, and it became exclusively an expression for North or the northerly direction. Early European maps called North America *America Septentriones,* literally "America of the Seven Plough-Oxen." In astrology, the first *six* signs of the zodiac were rather confusingly designated by the signs of the *seven* oxen, the "septentrional signs."

John Milton in *Paradise Regained* uses the term, again purely as an adjective for the northerly direction:

[B]acked with a ridge of hills
That screened the fruits of the earth and seats of men
From cold Septentrion blasts.

Shakespeare puts a similar expression in the mouth of the Duke of Gloucester in *Henry VI, Part III*:

Thou art as opposite to every good
As the Antipodes are unto us,
Or as the south to the septentrion.

On the other hand, the Bard has Sir John Falstaff simply use the number of Dipper lights for baser purposes in *Henry IV, Part I*:

[W]e that take purses
Go by the moon and the seven stars.

Not too different is the guidance given in John Keats's "Robin Hood": "the seven stars to light you/or the polar ray to right you."

Case 4B: Pleasant Sevens

Fasten your seat belt; "doing the sevens" will (and should) go by very quickly. Again, some of these "sevens" invoked by some speakers or writers may instead refer to the Pleiades.

Worthies: Boon companions of the sky have included the Seven Cabrillos (children) of Spain; the Seven Wise Men of Greece; the Seven Sleepers of Ephesus; the *Seitsen* *tahtiennen* of the Finnish epic poem *Kalevala;* the Seven Flammas of the Portuguese, appearing in the works of Camões; the *Yidighan Vildux* of the Turks; the Seven Champions of Christendom; the Seven Khans or *Jetiqaraqshi* of the Kazakhs; and the Seven Directors of the Chinese. In Hawaiian they are the Seven of Makalii, *Na Hiku ka Huihui a Makalii.*

Wise Men: Austere venerables of the heavens include the *Bintang Kartika*, the Seven Sages of the Javans; the Seven Anchorites (or monks) raised to the sky; and the Seven *Krttka* of Sanskrit. The latter are also known as the Seven *Riši* or Great Sages of Hinduism, also known as the *Saptarshi Mandala*, or *Saptar Shayar.*

Divinities: In Mongolian lore they have been the Seven Gods (as well as the Seven Khans). In one Japanese telling, they are the *Kami Amenominakanushi*, seven powerful spirits worshipped in Shinto. In Kalmyk, they have been reported as the seven wise gods.

Animals: The individual stars have been the Seven Bulls; the Seven Bears; the Seven Foxes; the Seven Elk; and the Seven Antelopes. Stories have been collected from the Puget Sound Snohomish of four elk and three hunters; from the Wasko-Wishram of five wolves and two bears left in the sky by the trickster Coyote; and from the Chumash of California of seven boys turned into geese—perhaps a cautionary tale against straying too far from one's village.

Brothers or Sisters: Several groups in central Asia are reported to see seven brothers; in this tradition, the diminutive Alcor is a girl or a child. The seven-brother story is also told by Turco-Tatars, Daghestanis, Armenians, and peoples of the Caucasus. A later Greek tradition sees seven brothers, or six brothers with a girl rather than a ruling child.

One Kurdish story also has seven brothers, but they were the stars of Ursa Minor instead! A Hindu tradition extends the story across two constellations: seven of the Pleiades are wives of the seven sages that are the Dipper stars.

In Korea, the stars have been referred to as the Seven Stars of the North. One tale is that a widow with seven sons sought the company of a widower who lived across a raging river. The sons covertly laid stepping-stones to help her across. Without knowing who placed the rocks, the widow asked the deities to bless the stone-layers. When the sons died, they ascended to heaven as the asterism.

Case 4C: Unpleasant Sevens

Then there are traditions that showcase unsavory characters and nasty situations in the heavens. As is often the case, some of these villains are more memorable than the preceding austere worthies, monks and sages.

Abductors: Iranian Ossetes are reported to see seven brothers who have stolen a child from the Pleiades, namely Alcor. Caucasus tribes also told of a reversal: it is the Pleiades that are scheming to take the girl Alcor from the Dipper, but have not done so yet. Other traditions where diminutive Alcor is seen as a sister abducted from the

Pleiades reportedly have included *(taking a deep stage-breath)* the Ancient Greeks, Bulgarians, Kumyks, the Nogai, Kazakhs, the Kirghiz, Altais, Tuvinians, Khakas, Altai Tatars, Hunter Uralics, Lapps, and the Yenisei. Many of these lie along part of the Mongol route toward Europe in the thirteenth century. In fact, the Milky Way in Russia is sometimes called the Road of Batyj, named after the grandson of Genghis Khan.

Combatants: There is a report of a Kabardines tale of seven brothers to whom their mighty Khan sent the child Alcor, to be raised by them to grow up to be the ruler (much like King Arthur). The Pleiades tried to abduct or kill the child, but the Dipper brothers returned at the last minute to defend him (much like the dwarves protecting Snow White).

Murder Victims: Buryat people are reported to have a grisly tale of seven brothers. The stars represent the upper part of the skulls of seven sons of a blacksmith. The Khan killed the sons and gave the skulls as wine cups to his wife, who threw the cups into the heavens after their contents had been drained.

Case 4D: The Parallelogram of Sevens

As can be seen, the Seven Brothers roll-call appears in several tribes in southern Siberia, the Caucasus and Mongolia. It likewise appears in certain tribes of Native Americans, particularly in what William Gibbon has charted as a "parallelogram" stretching from southwest Saskatchewan into southeast Oklahoma. That is correct, a parallelogram.

The Seven Brothers are apparently not found in the far north, or in the east, or in the west of North America. Gibbon's theory is that the Seven Brothers tradition came with those Eurasian migrants who populated this parallelogram, while those from the Bear or Elk traditions elsewhere in Eurasia populated North America outside the parallelogram.

Gibbon reckoned that ten out of fourteen groups of tribes in the parallelogram refer to the Seven Brothers. He cited the following evidence:

- The Crow, Wichita, and Cheyenne have a story of seven brothers living with the girl Alcor, who has fled her pursuer (a bear?). She is fearfully hiding behind one of the brothers, Mizar.

- The Blackfoot, Wichita and other Sioux tribes have stories telling of seven brothers and one or two girls here on Earth. Sometimes a sister remains on our planet as the Bear, while the seven brothers and the other girl go into the sky and become the 7+1 of the Big Dipper. Alcor may be either a wife or a child.

- The Kiowa are reported to have an opposite story, of seven sisters and one brother. The Seneca are reported to have a story of seven brothers, one of whom is carrying a weak brother, Alcor. A second brother leads with a torch, and a third is right behind Mizar and Alcor bearing a kettle.

- The Assiniboine of Winnepeg are reported to have perhaps the most gruesome of the stories. A father decapitates the mother, and the mother's severed head chases six sons and a daughter around the Earth. The seven children play ball in a circle. While their mother's head is still pursuing them, they ascend into the sky.

Ethno-astronomers and archaeo-astronomers have looked for ways to test Gibbon's theory. They have examined genetics, stories and historical evidence to consider whether tribes within the parallelogram can be traced back to groups remaining in Eurasia who also have seven-sibling traditions for this asterism.

Case 4E: The Ends of the Horizon

To conclude this roll-call category, I give you the Teda people of the African country of Chad. There, these seven stars regularly dip below the horizon. On the occasions where they do rise, the septet is sometimes known as wild asses.

Make of those asses what you will. I am sick of sevens.

HUMANITY AND THE BIG DIPPER

Figure 15. As Clouds, So Stars

HUMANITY AND THE BIG DIPPER

PATTERN CATEGORY 5: *Potpourri*

Recall that Homer knew that one person's bear is another person's wagon. The *Peanuts* comic *(see Figure 15)* illustrates that there is always someone who interprets nature very differently than the rest of us. Douglas Hofstadter cites this strip as an example of the "horses-and-doggies" syndrome, where one person makes an arcane association while the rest of us see something simple.

Several texts remind us that this circumpolar constellation rotates in front of our eyes.

- Back to near my beginning, the Greek poet Aratus sang of the pattern as *Helike,* the "twister"—because overnight and throughout the year, it swings around and around the pole.

- In a confusing mash-up, the Roman writer Lucan called the constellation *Pharsalian Helice*—*Helice* since it rotates and *Pharsalia* since the principal of the bear story, Kallisto, is said to have come from Pharsalia in the region of Arcadia.

- Some of the etymologies get rather messy. The Swedish researcher Peter Blomberg has suggested that the whole bear/north etymology of other scholars, as described above, is wrong—and that a Proto-Indo-European root, *hr̥tko-s or *hret meaning "roll" or "turn," is the actual source for the star names. In support, he cites the ancient Akkadian word *ereqqu* for "wagon." The matter seems far from settled, and I leave the controversy at that.

- The Navajos or Diné people are said to have referred to the Dipper as the Rotating Man, erect and one-legged, while the recumbent and enfolded *W* or *M* of nearby Cassiopeia forms the Rotating Woman. (I prefer to poeticize these translations as the "Turning Man" and the "Turning Woman.")

Alabama Native Americans are said to have called the entire pattern a Canoe. In Indonesia, the asterism has been completely independently called *Bintang Biduk*, the Canoe Stars.

Some south Indian groups, as reported by a French scholar, called the seven stars a Bed with one of the four legs broken off. The same scholar states that Athabascans described a One-Legged Man. Inuit peoples are said to see the stars as poles with skin ropes tied to them.

In ancient Japan, a Myoken family of beliefs arose associated with the two Bears, and with the North Star and the Big Dipper in particular. Myoken was first associated with several divinities. Later, it was blended with the seven *kami* tradition mentioned above. There are also references to a sword of seven stars (*schichi sei ken*). The *kami* are themselves associated with the physical layout of locations of seven shrines at Hie near Mt. Hieu.

Southern Chinese cultures are said to have used the constellation and its individually named stars in compilation of a detailed calendar. This calendar facilitated timekeeping in the form of the "Measurement" or the "Pacing" of Heaven, in some cases physically conducted on Earth by actual footwork along a floor diagram laid out as the Dipper stars themselves.

In northern China, the seven lights were sometimes referred to as the Government or the Bureaucrat (*Tseih Sing*), and the diminutive Alcor was dubbed the Supporting Star (*Goo Sing*). The four stars formed a box and were at times regarded as a single throne or multiple thrones of divinities, all of which were distinguished from the three stars of the tail, who were petitioners. Many more Chinese traditions accompany these stars.

A very different use of the stars was not as points of light but as a continuous boundary between regions of the darkness. In ancient Greece, Heraclitus noted that the pattern forms the boundary between East and West. In fact the Big Dipper does just that, especially when it is lowest in the heavens. Stansbury Hagar remarked, "The Chinese say that in spring the tail of the bear points east; in summer, south; in autumn, west; in winter, north—a correct statement for the forepart of the evening." In an old Chinese tradition, the Dipper protects the north pole as the Right Wall of the Purple Forbidden Enclosure (*Zeweigong*).

"The Right Wall of the Purple Forbidden Enclosure." Now *that* is a most impressive name. On this profound note, let us draw the curtain on this pageant of patterns.

.

EPILOGUE

As exhaustive and exhausting as this book's compilation must seem, I have not come close to touching bottom. There are tales of Zoroastrian generals, the Cities of Madrid and Antwerp coats of arms, the "Seven Sleeping Boys of Efesos," Chinese divinities, Icelandic sagas, alternative Greek myths, and more Native American animal stories, some featuring those infamous tricksters the Coyote and the Crow. The Big Dipper star Mizar appear in a Steely Dan song. Some of the patterns seen in Chinese, Japanese, Egyptian and Native American cultures are so integrated with religious traditions that I felt it best to stop respectfully at the initial stage of pattern recognition. Moreover, all my researches have been in the English language, and I have relied on written materials easily accessible to an online amateur. Further traditions must lie outside these covers.

Any essay of this type needs to end before it becomes an obsession and its author an Ahab. I am aware from viewing *Seinfeld* that this work bears a striking resemblance to Cosmo Kramer's coffee-table book entirely about coffee tables. Of all characters in English novels, I was perhaps most sympathetic to Edward Casaubon, the pedantic and ineffectual scholar in George Eliot's *Middlemarch*—not because he is a suitable husband for Dorothea Brooke or a worthy competitor for Will Ladislaw, but because I rooted for him to complete his unfinished and un-finishable masterpiece, *The Key to All Mythologies*. Eddie should have bitten off a more confined subject, like the Big Dipper.

It is enticing to survey the journey we have made. Beginning with a sea-urchin amulet and cave paintings of the Old Stone Age, we have traced the tellings about this constellation in Proto-Indo-European culture; observed as the sagas spread through migrations, conquest and trade throughout Eurasia and into the New World; followed the transmission of ideas from ancient civilizations in Egypt, Mesopotamia, Persia, Phoenicia, China, India, Palestine, Greece, and Rome to the Jewish, Christian and Muslim cultures of a first new millennium CE; and watched the passage of myths and writings into the giants of scientific and imaginative literature of a second millennium, and now a third. The story spans Europe, Asia, North America, Oceania and Africa. The entire northern hemisphere and some reaches well beyond have long been vibrant with songs of the seven.

But I also resist the temptation to think of a grand parade of peoples who are connected through blood, travel or wars. All of us are heirs of all the stories of all of humanity—not just those of our tribe and our time, but of every tribe and every time. For that reason, throughout this study I have de-emphasized the genetic, ethnic and historic aspects, and focused on the patterns themselves. These are all products of the imagination that is our common faculty. I emerge in awe of the inventiveness and richness of our time on this planet to date.

When any humans have looked at the fixed lights of the Big Dipper, in reality they have seen a spectacular and ongoing set of hydrogen-fusing-into-helium explosions that took place 80 to 123 years before. Nothing remains constant forever. Modern physics has transported our entire worldview from a universe composed of stable matter to one where all is in the process of becoming.

As physicist Carlo Rovelli summarized, "the world is made of events, not things." Fundamental particles and their opposites flicker into and out of existence; energy and mass both seen and unseen convert into one another; and the coordinates of space and time depend on the situation of the observer. On the timescale of the cosmos, our life on Earth is like a passing film of lichen on a briefly heated boulder. For that very reason, it is how we live and how we care for others that ultimately matter.

Two thousand years from today, Polaris will have been replaced as the north star by a new title-holder. Sometime after that, the inner star cluster will separate from the two outer stars, and the Big Dipper will have been torn asunder. The old stories will lose their footing; the old patterns will no longer have purchase.

Against all evidence to the contrary in our recent history, let us assume that our species survives for many millennia, in some form, whether fleshly or electronic. What new patterns will we make and what new stories will we tell about our new night sky? And what will happen to our stories and patterns if—and when—we humans or our successors slip the surly bonds of Earth and inhabit other worlds?

I conclude with a tale not of the past but of the future, not about the Big Dipper but about an adjacent group. We started by invoking the help of a muse, and Goethe ended his *Faust* with a mystical chorus singing of *Ewig-Weibliche,* the Eternal Feminine. Let us end this journey by shifting our focus from the Turning Man of the Navajos to his enduring counterpart the Turning Woman, the nearby northern constellation Cassiopeia.

Cassiopeia is highlighted by a *W*- or *M*-shaped asterism. It lies directly opposite on the celestial sphere from the southern constellation Centaurus. As is well known, Alpha Centauri and its companions are the closest stars to our own solar system, and thus the most likely first destinations of interstellar travelers from Earth.

Our progeny making that journey will undoubtedly look back toward our Sun from their new home, several light-years away. What will they see?

They will observe one new extra zigzag transforming the *W* or *M* of Cassiopeia into a different constellation—a jagged "Sawtooth," perhaps. Our great-grandchildren of some remove will spot our home fire making its debut as a fairly bright star—one with 0.5 apparent magnitude, comparable to the Little Dog's leading light Procyon, the kid brother unto Sirius.

Beholding that pattern, enfolded and recumbent, under what I hope are cloudless climes that distant alien evening, our descendants might recall a story not of Homer or Shakespeare, but of a younger bard in a latter day:

Then felt I like some watcher of the skies
When a new planet swims into his ken.
   —JOHN KEATS

∞ ∞ ∞

HUMANITY AND THE BIG DIPPER

.

ACKNOWLEDGMENTS

In the realms of astronomy and anthropology, I travel with the tourist's visa of a layperson. I have discovered no sextuple stars, interviewed no elderly members of a subgroup, unearthed no archaeological sites, and newly translated no original sources. As noted in the text, I have only consulted English-language sources, and even those have been confined to texts readily available to me.

I am therefore grateful and indebted to those who have produced the primary and secondary sources cited in the Bibliography. Listing of one authority in a Note is only for minimum reference, and other sources often could have been cited for the same proposition. Anyone with an interest in the physical sciences, the social sciences and the humanities inherent in this survey is invited to access the works of the professionals themselves.

I am grateful to Joel Bresler, a fellow amateur in this field who, much like me, was taken aback when he learned there was much more to a story than what he had been told. He was motivated to seek, digest and communicate the facts that he could garner, and he showed me the way by his evenhanded, cautious, and civil approach.

And to the families to whom I told those Big Dipper stories, I am sorry, and offer this book as my apology.

HUMANITY AND THE BIG DIPPER

NOTES

PROLOGUE

Homer epigram: Iliad, book XVI (Scott Dickerson tr.).

Equality of Dipper stars: Manilius, quoted in Bouvier.

Van Gogh letter: Vincent Van Gogh to Theo Van Gogh, Sept. 29, 1888.

Homer direction: Odyssey, book V.

WHAT OUR SENSES SEE

Polaris and constellation facts: Ridpath & Tirion; Ridpath.

WHAT THERE REALLY IS

Alcor-Mizar sextuple: Mamajek et al.

Alcor: Berezkin 2005.

Vision test: Allen, Gore, Ridpath.

Moving Group: Schaefer.

WHAT OUR MINDS CREATE
Psychological origins of constellations: Jung, Campbell, Hofstadter, Ackerman, Koffka.
Historic and ethnic origins of constellations: d'Huy, d'Huy & Berezkin.
Finn and Numic nets: Allen, d'Huy 2013, Berezkin 2005.
Apollo 8 Earthrise photo: Brand; Drake; George.

IS THERE A GREATER BEAR?
Primitive Bear culture: Frank 2014, Frank 2016, Antonello, Hollowell.
Skeptics of a bear shape: Schaefer, Collins, Rogers II, Frank 2014, Ruggles, Olcott, L'Engle.
Backwards-facing bear: H. A. Rey.

1: THE SCENE
Ancient Sky Bear: Schaefer, Olcott.
Four-star Bear: Schaefer, Berezkin 2005, d'Huy 2013.
Orion or Boötes Bear: d'Huy & Berezkin.
Dipper amulet: Baudouin, Makemson.
Origin of Bear story for Eurasia and the Americas: Schaefer.
Biblical scholars: Henry.
North American Cosmic Hunt: Hagar, Gibbon, Allen.
Criticisms of Hagar: Eddy, Thompson.
Cosmic Hunt variants: Children's tale, Monroe & Williamson; North American variants, Hagar, Allen; Eurasian variants, Berezkin 2005, d'Huy 2013, Gibbon, Olcott, Schaefer.
Significance of Cosmic Bear Story: Krupp 1991.
Coeur d'Alene grizzly bear story: Gibbon.
Funeral bier: Olcott.

Book of Job: Bible Vulgate, Bible King James version, *Book of Job* (Alter tr.), Strong's *Concordance.*
Arabian scenes: Rogers I.
Christian scenes: Allen.
Koryak scene: Gibbon.
North American scenes: Thompson.
Bed and table: Berezkin 2005.

2: THE ANIMAL

Proto-Indo-European: Blomberg.
Aristotle: Meteorologica II, c. 5, 362^b 8-9, 363^b 15.
Strabo: Geographica I, c. 1.
Greek terms: Olcott, Allen.
Thales: Schaefer, Diogenes Laertius, *Lives of the Eminent Philosophers* 1, "Thales" (citing Callimachus).
Erastosthenes: Condos.
Ptolemy of Alexandria: Systeme Mathematike, Almagest.
Strabo, Geographica.
Ovid. Metamorphoses, book II, Fabula VI.
Vergil: Aeneid, book III.
Dante: *Paradiso*, canto II, stanza 1.
Hebrew and Arabian Bears: Allen.
European, Sanskrit Bears: Allen, Olcott (citing Max Müller), Gibbon, Camões, García Lorca, "Canción para la luna."
North American Bears: Gibbon, Hagar, Thompson.
Shakespeare: Othello, act II, scene 1.
Elk: d'Huy, Lushinilova 2003, Antonello, Gibbon.
Russian elk/cart line: Berezkin 2009.
Arabian animals: Allen, d'Huy 2013.
Lapps, Inuit: Allen, MacDonald.

Minoans: Blomberg.

Paiutes and Hidatsa: Dorcas Miller.

Ojibwas: Schoolcraft.

Aztec and Mayan: Makemson, Krupp, Tedlock & Tedlock.

Egypt: Krupp, Allen, Olcott, Antonello.

Birds: Allen, Berezkin 2005.

Scorpion: Gibbon.

Circumpolarity disadvantage: Berezkin 2009.

3: THE IMPLEMENT

Wheel: Schaefer.

Akkadian: Rogers I.

Later Assyrian: Bradley, Schaefer, Rogers I.

Transmission to Thales: Allen.

Latin Septentrion: Allen.

Latin Plaustrum and European Wagons: Germanicus Caesar, Allen.

Chinese Chariot: Krupp 1991.

OED: Oxford English Dictionary, entries for "Wagon," "Wain," "Charles's Wain."

Coverdale Bible (1535).

Spenser: The Faerie Queene, book 4.

Shakespeare: Henry IV Part I, act II, scene 1.

Tennyson: "New-Year's Eve," "In Memoriam," "The Princess."

Plough shape: Germanicus Caesar, Aratus, Ridpath, Propertius, Allen.

India yoke: Allen.

"Plough" not found in USA: Merriam-Webster's Eleventh New Collegiate Dictionary.

"Dipper" definitions: Webster's Dictionary 1828; Merriam-Webster International Dictionary 1890.

Little Dipper, other American and European utensils: Allen.
Asian utensils: New World Encyclopedia.
Tolkien sickle: Tolkien, Larsen.
Gourd tale: Hockey, New World Encyclopedia, Berezkin 2005, Wilder (citing Winter).
Gourd tale cultural history: Bresler, followthedrinkinggourd.org.
Underground Railroad memoirs: Adams, Still.
"Gourd" not a published pre-1928 US term: Allen, Olcott (quoted), Park.
Gourd song story discounted: Kelley, Henry Louis Gates, Jr., Scholastic, Riddle.

4: THE ROLL-CALL
"Separate persons": Schaefer.
Power of lucky/ magic seven: "Seven Reasons" *Psychology Today* article, Kolb.
Seven Stars pubs: Kolb.
Septentrion and seven in literature: Hyginus; Dante, *Purgatorio,* canto XXX; John Milton, *Paradise Regained,* book 4; Shakespeare, *Henry VI Part III,* act I, scene 4; Shakespeare, *Henry IV Part I,* act I, scene 2; John Keats, "Robin Hood."
Pleasant and Unpleasant Sevens: Allen, Makemson, d'Huy 2012, Berezkin 2005.
Parallelogram of Sevens: Gibbon.
Sioux sevens: Hungry-Wolf.
Seven Asses of Chad: Berezkin 2009.

5: POTPOURRI
Peanuts *comic:* Hofstadter.
Lucan, Pharsalia, book II.

Navajo Turning Man and Woman: Chamberlain.
Etymology: Blomberg.
Canoe, Bed, One-legged man, poles: Berezkin 2005, d'Huy 2013.
Japanese terms: Arichi, Heritage of Japan.
Chinese terms: Allen, Hagar, Penprase, Makemson.

EPILOGUE
Cassiopeia/Centauri: LearnAstronomyHQ.
New planet: Keats, "On First Looking Into Chapman's Homer."

BIBLIOGRAPHY

Asger Aaboe, "What every young person ought to know about naked-eye astronomy" (1976), reprinted in *Episodes in the Early History of Astronomy* (2001).

Diane Ackerman, *A Natural History of Love* (1995).

Samuel Hopkins Adams, *Grandfather Stories* (1955).

Dante Alighieri, *Divine Comedy* (Henry Wadsworth Longfellow tr. 1867).

Richard Hinckley Allen, *Star-Names and Their Meanings* (1899).

Robert Alter, tr. *The Wisdom Books: Job, Proverbs and Ecclesiastes* (2011).

Elio Antonello, *The Myths of the Bear* (2011).

Meri Arichi, "Seven Stars of Heaven and Seven Shrines on Earth: The Big Dipper and the Hie Shrines in the Medieval Period," 10 *Culture and Cosmos* 195 (2006).

Aratus, *Phaenomena* (Robin Hard tr. 2015).

Aristotle, *Meteorologica* (E. W. Webster tr. 1923).

Matthew Arnold, *Merope: A Tragedy* (1858).

Marcel Baudouin, "La Grande Ourse et le Phallus du Ciel," 18 *Bulletin de la Societé préhistorique française* 11 (1921), p. 301.

John C. Barentine, *Uncharted Constellations* (2016).

Johann Bayer, *Uranometria* (1603), available at http://lhldigital.lindahall.org/cdm/ref/collection/astro_at las/id/118.

Yuri Berezkin, "The Cosmic Hunt: Variants of a Siberian-North-American Myth," 31 *Folklore* 79 (2005).

Yuri Berezkin, "Seven Brothers and the Cosmic Hunt: The European Sky in the Past," in *Universumit Uudistades Paar Sammukest XXVI. Eesti Kirkjanausmuseum Aastaraamat* 31 (2009).

The Bible, Vulgate Latin translation (432), Coverdale English translation (1535), Douai-Reims Vulgate English translation (1610), King James version (1611).

J.F. Blake, *Astronomical Myths* (1877).

Peter E. Blomberg, "How Did the Constellation of the Bear Receive Its Name?" *Archaeastronomy in Archaeology and Ethnology* 129 (n.d.).

Peter E. Blomberg, "The Northernmost Constellations in Early Greek Tradition" (n.d.)

Hannah M. Bouvier, *Bouvier's Familiar Astronomy* (1857).

Stewart Brand, *The Whole Earth Catalog* (2d ed. 1969).

Joel S. Bresler, "Follow the Drinking Gourd: A Cultural History," at http://followthedrinkinggourd.org.

Luis de Camões, *The Lusiads* (W. J. Mickle tr. 1877).

Joseph Campbell, *The Way of the Animal Powers* (1983).

Chris Cannon & Gary Holton, "A Newly Documented Whole-Sky Circumpolar Constellation in Alaskan Gwich'in," 51:2 *Arctic Anthropology* 1 (2014).

Von Del Chamberlain, John B. Carlson & M. Jane Young (eds.), *Songs from the Sky: Indigenous Astronomical and Cosmological Traditions of the World* (2005).

Geoffrey Chaucer, *Boethius* (Richard Morris ed. 1868) and *Canterbury Tales* (Walter W. Skeat ed., 2d ed. 1900).

M. Tullius Cicero, *Aratea* and *De oratorio* (W. A. Falconer tr. 1923).

A. Frederick Collins, *The Book of Stars* (1920).

Theony Condos, "The *Katasterismoi* [of pseudo-Eratosthenes]," *Bulletin of the Astronomical Society of the Pacific* (October & November 1970).

John Davies, *Poems* (1626).

Julien d'Huy, "Un ours dans les étoiles: recherché phyolgénétique sur un mythe Préhistorique" (English abstract), 20 *Bulletin Préhistore du Sud-Ouest* 91 (2012).

Julien d'Huy, "A Cosmic Hunt in the Berber Sky," 16 *Les Cahiers de l'AARS* 93 (2013).

Julien d'Huy, "The Evolution of Myths," *Scientific American* 62 (December 2016).

Julien d'Huy & Yuri Berezkin, "How did the first humans perceive the starry night? On the Pleiades," *Retrospective Methods Network Newsletter,* University of Helsinki (2017).

Diogenes Laertius, *Lives of the Eminent Philosophers,* volume 1 (Thales) (R.D. Hicks tr. available at http://www.perseus.tufts.edu/hopper/).

Nadia Drake, "We Saw Earth Rise Over the Moon in 1968. It Changed Everything," *National Geographic* (Dec. 21, 2018).

John Eddy, "Archaeoastronomy of North America," in E.C. Krupp ed., *In Search of Ancient Astronomies* (1978).

Encyclopedia of Religion and Ethics, vol. XII, "Sun, Moon and Stars (Iranian)" (1958).

Eratosthenes or followers, attributed summaries of *Catasterisms* (Robin Hard tr. 2015).

Thomas Fale, *Horologiographica* (1593).

Roslyn M. Frank, "Origins of the 'Western' Constellations," in Clive L.N. Ruggles, *Handbook of Archaeoastronomy & Ethnoastronomy* vol. I, p. 214 (2014).

Roslyn M. Frank, "Sky Bear Research: Implications for 'Cultural Astronomy,'" 16 *Mediterranean Archaeology & Archaeometry* 343 (2016).

Federico García Lorca, *Selected Poems* (1941).

Henry Louis Gates, Jr., "Who Really Ran the Underground Railroad?" available at pbs.org/wnet/african-americans-many-rivers-to-cross/.

Alice George, "How Apollo 8 Saved '1968,'" *Smithsonian* (December 11, 2018).

Germanicus Caesar, attributed translation of *Phaenomena* by Aratus (D. B. Gain tr. 1971).

William B. Gibbon, "Popular Star Names among the Slavic-Speaking Peoples," Ph.D. dissertation, University of Pennsylvania (1960).

William B. Gibbon, "Asiatic Parallels in North American Star Lore: Ursa Major," 77 *Folklore* 236 (July-Sept. 1964).

William B. Gibbon, "Asiatic Parallels in North American Star Lore: Milky Way, Pleiades, Orion," 85 *Folklore* 236 (July-Sept. 1972).

Owen Gingerich, "Astronomical Scrapbook: Origin of the Zodiac," 67 *Sky & Telescope* 218 (March 1984).

J. Ellard Gore, "The Names of the Stars," 286 *Gentlemen's Magazine* (January 1899) p. 17.

Alexander A. Gurshtein, "The Origins of the Constellations," 85 *American Scientist* 264 (1997).

Stansbury Hagar, "The Celestial Bear," 13 *Journal of American Folk-Lore* 92 (1900).

Nathaniel Hawthorne, *Passages from French & Italian Note-Books* (1871).

Jonathan F. Henry, "Constellations: Legacy of the Dispersion from Babel," 22 *Journal of Creation* 93 (2008).

Heritage of Japan, "Big Dipper cult and Myoken worship in Japan" (Jan. 22, 2013), available online.

Hesiod, *Astronomia* and *Eoeae* (Categories of Women) fragments (Glenn C. Most tr. 2007).

Hipparchus, *Commentary on the Phenomena of Aratus and Eudoxus* (Karl Manitius ed. 1894) (Latin).

Thomas Hockey, *How We See the Sky: A Naked-Eye Tour of Day and Night* (2011).

Douglas R. Hofstadter, *I Am a Strange Loop* (2007).

J. C. Holbrook & Audra Balesis, "Naked-Eye Astronomy for Cultural Astronomers," in *African Cultural Astronomy* (2008).

Homer, *Iliad* (Robert Fagles tr. 1990).

Homer, *Odyssey* (Robert Fagles tr. 1996).

Thomas Hood, *The Use of the Celestial Globe in Plano* (1590).

Adolf Hungry-Wolf, *Blackfoot People* (2006).

Lewis Hyde, *Trickster Makes This World: How Disruptive Imagination Creates Culture* (1999).

G. J. Hyginus, *Astronomy* (Robin Hard tr. 2015).

Samuel Johnson, *A Dictionary of the English Language* (1755).

Carl Gustaf Jung, *The Archetypes and the Collective Unconscious* (1968).

John Keats, *Complete Poems* (John Barnard ed. 1977).

James B. Kelley, "Song, Story or History: Resisting Claims of a Coded Message in the African-American Spiritual 'Follow the Drinking Gourd,'" 41 *Journal of Popular Culture* 262 (2008).

K. Koffka, *Principles of Gestalt Psychology* (1954).

Hugh Kolb, *Seven Stars: Ancient Astronomy and the English Public House* (2019).

E. C. Krupp, *Beyond the Blue Horizon: Myths and Legends of the Sun, Moon, Stars and Planets* (1991).

E. C. Krupp, *Echoes of the Ancient Skies: The Astronomy of Lost Civilizations* (2d ed. 1994).

E. C. Krupp & Robin Rector Krupp, *The Big Dipper and You* (1989).

George E. Lankford, *Reachable Stars: Patterns in the Ethnoastronomy of Eastern North America* (2007).

Kristine Larsen, "Myth, Milky Way, and the Mysteries of Tolkien's *Morwinyon, Telumendil,* and *Anarríma,*" 7 *Tolkien Studies* 197 (2010).

LearnAstronomyHQ, "How Would Our Sun Look From Our Nearest Star System?" available at https://www.learnastronomyhq.com/articles/how-would-our-sun-look-from-alpha-centauri.html.

Madeleine L'Engle, *An Acceptable Time* (1989).

M. Annaeus Lucan, *The Pharsalia of Lucan* (Edward Riddley tr. 1896).

John MacDonald, *Arctic Sky* (1998).

Maud Makemson, "Astronomy in Primitive Religion," 22 *Journal of Bible & Religion* 163 (1954).

Linda A. Malcor, "The Icelandic Sword in the Stone: Bears in the Sky," *The Heroic Age* (May 2008).

M. Manilius, *Astronomicon* (Thomas Creech tr. 1700).

Eric E. Mamajek et al. "Discovery of a Faint Companion to Alcor," 139 *Astronomical Journal* 919 (2010).

G. & C. Merriam Co., *Webster's International Dictionary* (1890).

Merriam-Webster's Eleventh New Collegiate Dictionary (2003).

Dorcas S. Miller, *Stars of the First People: Native American Star Myths and Constellations* (1997).

John Milton, *Paradise Regained* (1671).

Jean Guard Monroe & Ray A. Williamson, *They Dance in the Sky: Native American Star Myths* (1987).

New World Encyclopedia contributors, "Big Dipper," *New World Encyclopedia* (June 7, 2016), https://www.newworldencyclopedia.org/p/index.php?title=Big_Dipper&oldid=996565 (accessed March 16, 2020).

William Tyler Olcott, *Star Lore of All Ages* (1911).

Michael Ovenden, "Origins of the Constellations," 3 *Philosophy Journal* 1 (1966).

P. Ovidius Naso (Ovid), *Metamorphoses* and *Fasti* (Henry T. Riley tr. 1893).

Oxford English Dictionary (2d ed. 1989), entries on "wagon," "wain," "Charles Wain," "plough," "septentrion," "dipper."

William R. Palmer, *Why the North Star Stands Still and Other Indian Legends* (1957).

William Peck, *The Constellations and How to Find Them* (1884).

Bryan E. Penprase, *The Power of Stars* (2d ed. 2017).

Sextus Propertius, *Elegies*, Elegy II (P J. F. Gantillon tr. 1895).

Ptolemy of Alexandria, *Almagest,* books VII and VIII (G. J. Toomer tr., Owen Gingerich ed. 1999).

Gloria D. Rall, "The Stars of Freedom," *Sky & Telescope* (February 1995).

H. A. Rey, *The Stars: A New Way to See Them* (1952).

Bob Riddle, "Scope on the Skies," *ScienceScope* (Feb. 2006) 54.

Ian Ridpath, *Star Tales* (rev. ed. 2018).

Ian Ridpath & Wil Tirion, *Stars and Planets* (2017).

John H. Rogers, "Origins of the ancient constellations: I. The Mesopotamian traditions," 108 *Journal of the British Astronomical Association* 9 (1998) ("Rogers I").

John H. Rogers, "Origins of the ancient constellations: II. The Mediterranean traditions," 108 *Journal of the British Astronomical Association* 49 (1998) ("Rogers II").

Carlo Rovelli, *The Order of Time* (Erica Segre & Simon Carnell trs. 2017).

Clive L. N. Ruggles, *Ancient Astronomy* (2005).

William Sale, "The Story of Callisto in Hesiod," *Rheinisches Museum für Philologie, Neue Folge,* 105 *Band* 2, R. 122 (English text) (1962).

Bradley E. Schaefer, "The Origin of the Greek Constellations," *Scientific American* 96 (November 2006).

Scholastic Inc., "Myths of the Underground Railroad," at teacher.scholastic.com/activities/bhistory/underground_railroad/myths.htm.

Henry R. Schoolcraft, *The Legend of Hiawatha* (1856).

Walter Scott, *Lay of the Last Minstrel* (1805).

"Seven Reasons We Like Seven Reasons," *Psychology Today* (June 27, 2015).

William Shakespeare, *The Norton Shakespeare* (Stephen Greenblatt et al. eds. 2015).

James R. Sowell, *The Naked-Eye Sky* (2d ed. 2011).

Edmund Spenser, *Edmund Spenser's Poetry* (4th ed. 2013).

Julius D. W. Staal, *The New Patterns in the Sky* (2d ed. 1996).

William Still, *The Underground Railroad: A Memoir* (1871).

Strabo, *Geographica,* volume 1 (H. L. Jones tr. 1969).

James Strong, *The Exhaustive Concordance of the Bible* (1891).

Dennis Tedlock & Barbara Tedlock, "A Mayan Reading of the Story of the Stars," *Archaeology* 33 (July/August 1993).

Alfred, Lord Tennyson, *Tennyson* (Christopher Ricks ed. 1989).

Gary D. Thompson, "Ancient Zodiacs, Star Names, and Constellations," http://members.westnet.com.au/gary-david-thompson/index1.html.

Henry David Thoreau, *Autumn: From the Journal of Henry David Thoreau* (H. G. O. Blake ed. 1892).

J. R. R. Tolkien, "Myths Transformed," in *Morgoth's Ring* (vol. X of *The History of Middle-Earth,* Christopher Tolkien ed. 1993).

P. Vergilius Maro (Vergil), *Aeneid* (John Dryden tr. 1697) and *Georgics* (T. F. Royds tr. 1907).

Vincent Van Gogh, *The Letters* (Robert Harrison tr. available at webexhibits.org/vangogh/letter/18/543.htm, permission to use subject to Creative Commons license).

Noah Webster, *Webster's Dictionary* (1828).

Wikipedia (English), entries on "Big Dipper," "Boötes," "Follow the Drinkin' Gourd," "Ursa Major," "Ursa Minor."

Christina Wilder, "Follow the Drinking Gourd: Bibliography," 89 *Proceedings of the Astronomical Society of the Pacific* 291 (1996).

Jeanette Winter, *Follow the Drinking Gourd* (1988).

Clark Wissler, *Star Legends Among the American Indians* (1936).

Robert Wrigley, "The Big Dipper," 9 *Missouri Review* no. 2, 204 (1986).

HUMANITY AND THE BIG DIPPER

HUMANITY AND THE BIG DIPPER

HUMANITY AND THE BIG DIPPER

HUMANITY AND THE BIG DIPPER

W